MACMILLAN WORK OUT SERIES

Waves and Optics

J. Beynon

M
MACMILLAN

First published 1988
Reprinted 1990

Published by
MACMILLAN EDUCATION LTD
Houndmills, Basingstoke, Hampshire RG21 2XS
and London
Companies and representatives
throughout the world

Typeset by TecSet Ltd,
Wallington, Surrey
Printed in Hong Kong

British Library Cataloguing in Publication Data
Beynon, J.
Work out waves and optics.—
(Macmillan work out series).
1. Optics
I. Title
535 QC355.2
ISBN 0–333–46334–X

For my wife Christine, my children Alison and Andrew, and my parents

Contents

Acknowledgements

The author and publishers wish to thank the following who have kindly given permission for the use of copyright material:

Addison Wesley Publishing Company for a question from *Optics* by F. W. Sears, 1964

Cambridge University Press for a question from *Cavendish Problems in Classical Physics* by A. B. Pippard, 1971

Bruno Rossi for questions from *Optics*, Addison-Wesley Publishing Company, 1957

Hilda F. Smith for a question from *Optics* by C. J. Smith, Edward Arnold, 1960

John Wiley & Sons Ltd for a table from *Introduction to Matrix Methods in Optics* by A. Gerrard and J. M. Burch, 1975

Department of Physics, Brunel University for questions that have appeared in past undergraduate examination papers

Every effort has been made to trace all the copyright holders but if any have been inadvertently overlooked the publishers will be pleased to make the necessary arrangement at the first opportunity.

Introduction

The aim of this book is to help you, the student, develop a strategy for solving problems of varying complexity. The book is written in the hope that it will form part of your overall study/revision programme in the general field of optics. Optics is a subject which has increased in importance in recent years through the advent of the laser and the development of optical-fibre technology and optical communication systems. Thus the reason for studying this subject in depth cannot be over-emphasised.

Unlike a number of other textbooks, the method used to solve the problems will not be to substitute numbers into an appropriate expression, apparently 'plucked out of the air', but, instead, to lead up to a solution through a discussion of the underlying principle, out of which relevant formula(e) can be extracted. Using this problem-solving methodology you should be able to tackle more advanced problems in optics as well as in other areas of physics and engineering.

Problem-solving Methodology

Before a problem can be solved it is essential:

(1) to recognise the particular topic being considered;
(2) to understand, in broad terms, the principle(s) and concept(s) that underpin the theory, and
(3) to search for relevant equations that may be needed for the solution.

The steps used to solve problems are:

(1) Identify the physical principle(s).
(2) Draw a meaningful diagram or sketch, well-labelled and incorporating relevant information.
(3) Extract given data from the question and list unknown parameters.
(4) Write down all relevant equations.
(5) Identify which equation(s) satisfy the given data.
(6) Obtain an algebraic solution, whenever possible.
(7) Substitute the given data into the algebraic solution.

Steps (1)–(7) may be looked upon as a checklist to help you through the problem-solving process. Even though you may decide not to write down each step as completely as in this book it would be wise to follow the same basic procedure, at least in the early stages whilst you are developing your own problem-solving strategy.

It is pertinent to mention at this point that step (1) discusses the basic concepts and background to the problem in a superficial way only. This book is not a textbook on optics. To get maximum benefit from this book it is advisable to use it in conjunction with one of the many standard textbooks and your lecture notes.

The diagram (step 2) is particularly important. It should be sufficiently large so that all essential information can be written on it without over-crowding. So often students draw diagrams which are small and untidy. It is a great help if the diagram is about one-third of a page in size, with clear labelling. There is no need for a technical drawing – a free-hand sketch is perfectly adequate.

Although it is particularly satisfying to obtain a correct numerical result to a problem, inevitably errors will occur. There is no need to worry too much about this because few marks ought to be lost in an examination, especially if you have managed to derive the appropriate algebraic expression.

The examples contained in this book may be either studied in a systematic way by working steadily through each chapter or in a random way by skipping from topic to topic. There is no best way of using this book although maximum benefit will almost certainly be derived if it accompanies a main optics textbook. There are a number of unworked examples at the end of each chapter, together with outline solutions. A number of multi-choice questions test your basic optics knowledge.

This book is designed to be of help to first- and second-year students of Pure Physics, Applied Physics or some Physics-based subject in university, polytechnic, institute of higher education or college of technology. The range of examples should satisfy all needs. Some school pupils of Physics should also find this book a useful preparation for their future science education.

It is anticipated that most students will have some basic knowledge of the calculus and matrices. However, the appendix is included as a brief refresher on matrices – it deals with the rules for manipulating 2×2 matrices; multiplication of 2×2 matrices is all that is required for carrying out the matrix problems in Chapter 1.

1 Geometrical Optics

1.1 Introduction

Geometrical optics is often said to be an alternative way to wave optics for describing the passage of light through a medium. In fact, more accurately, geometrical optics is an approximation to wave optics. Using Huyghens' Principle (Section 4.2) we find that a wavefront expands in all directions in free space (i.e. in the absence of apertures and obstacles) and that light waves propagate along directions which are normal to the wavefronts. A ray is an approximation to a light wave in free space and is identical with the wave normal. The electromagnetic theory of radiation goes further and says that a ray is the direction of the Poynting's vector of the radiation field, i.e. the direction along which energy is transported. In addition, light rays obey Fermat's Principle of Least Time in going from one point to another. This principle can be used to explain the origin of the mirage and the apparent position of the setting sun, for example. All the examples in this chapter are concerned with light rays in the *paraxial* or *Gaussian* domain. That is, the rays make a small angle with the optical axis of the system so that the sine and tangent of the angle can be replaced by the angle expressed in radians – at least up to about 15°.

1.2 Refractive Index

Snell's law of refraction says that when light is refracted at the interface between two media

$\sin i/\sin r$ is a constant (usually defined to be > 1) (1.1)

where i is the angle between the direction of the ray and the normal in the less optically dense medium and r is the corresponding angle in the denser medium. This constant is given the name of *refractive index*. It is written ${}_1n_2$. Although the refractive index can be defined for any two media it is usual to choose air or vacuum as a reference medium with a value of 1.0000. In fact, the refractive index of air depends on the pressure P and should accurately be expressed by the relation

$$n_A = 1.0000 + kP \quad (1.2)$$

where k is a sensitivity constant. However, in order to simplify the form of the single surface and thin and thick lens equations we shall always take n_A as 1, unless otherwise stated.

1.3 Sign Conventions

There are a number of sign conventions that can be adopted to aid solutions of the various problems dealing with refraction at a single curved surface and at thin and thick lenses – such as the 'Real-is-Positive' and the 'New Cartesian' systems.

The former sign convention will be followed in this book and the details are given below. It should not be difficult to follow them because they are used consistently in the solutions. Nor should it be difficult to transfer to some other sign convention with which you may be more familiar.

(a) Spherical Surface and Thin Lens

The rules of the sign convention are given for the case in which the light is incident on the optical system from left to right. These are:

(1) The object distance x_o is positive if measured to the left of a vertex.
(2) The image distance x is positive if measured to the right of a vertex.
(3) The focal length of a convex surface or a converging lens is positive.
(4) The focal length of a concave surface or a diverging lens is negative.
(5) A convex surface has a positive radius of curvature.
(6) A concave surface has a negative radius of curvature.
(7) Object and image sizes are positive if measured above the optical axis.

If, in any problem, the direction of the incident light is from right to left then (5) and (6) must be interchanged.

The relevant equation to use with the spherical surface is

$$n/x_o + n'/x = (n' - n)/r \qquad (1.3)$$

where n and n' are the refractive indices of the media on the two sides of the surface and r is its radius of curvature.

The equation to use with a thin lens separating two media of refractive indices n and n' is

$$n/x_o + n'/x = n/f = n'/f' \qquad (1.4)$$

where f and f', respectively, are the primary and secondary focal lengths. f is equal to f' if the media are identical.

(b) Thick Lens

In problems involving a thick lens you can no longer ignore the thickness of the lens. A knowledge of the six cardinal planes of the lens must be obtained before a complete analysis of the system is made. Although, if the medium on both sides of the lens is the same then the principal planes and nodal planes coincide. Once the positions of the cardinal planes have been determined the lens structure can be dispensed with. So for a thick lens in air Fig. 1.1 is required.

Once again it is essential to adopt a sensible sign convention. The following rules are used in all the thick lens (and lens combination) examples:

(1) The object distance x_o and primary focal length f are positive if measured to the left of the first principal plane P_1.
(2) The image distance x and the secondary focal length f' are positive if measured to the right of the second principal plane P_2.
(3) If a principal plane lies to the right of its respective vertex then the distance between them is said to be positive, i.e. AH_1 is positive and BH_2 is negative.
(4) Object and image sizes are positive if measured above the optical axis.

As we shall only be considering a thick lens in air, the object/image equation to be used is

$$1/x_o + 1/x = 1/f \qquad (1.5)$$

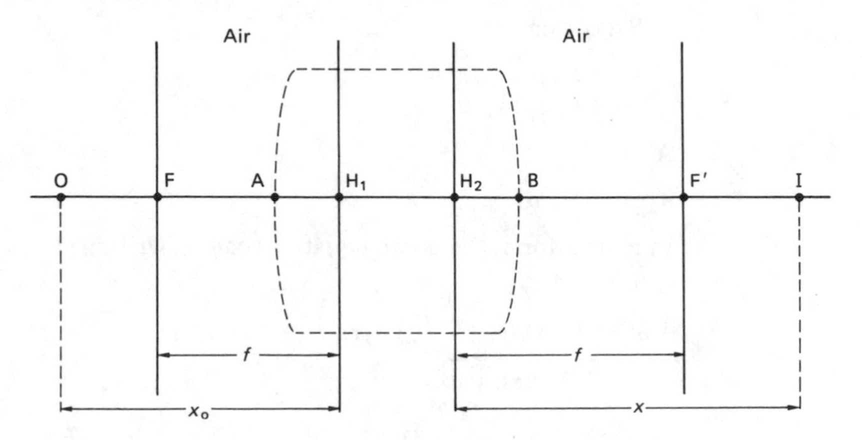

Figure 1.1

There is an alternative form of lens equation to (1.4) and (1.5) called the Newtonian representation. It says that if s_o is the displacement of the object from the primary focal point and s is the corresponding displacement from the secondary focal point then, in general,

$$s_o s = ff' \tag{1.6}$$

s_o and s are both positive if the object lies outside the primary focal point and the image outside the secondary focal point, as indicated in Fig. 1.1.

1.4 Matrix Representation

Matrices offer an alternative approach to the refracting surface/lens equations for following the path of a light ray through an optical system. Matrices are widely used in advanced lens design. Educationally, matrices offer some advantages over the conventional equations in that the process of translation through a medium and refraction at an interface must be clearly defined.

Consider Fig. 1.2 in which a ray leaves a point P, a distance z above the optical axis, and is directed upwards to meet a refracting surface at R. The angles α_1 and α_2 are both positive because the axis must be rotated in an anti-clockwise direction to become coincident with the ray. If the ray is directed downwards then the angle is negative.

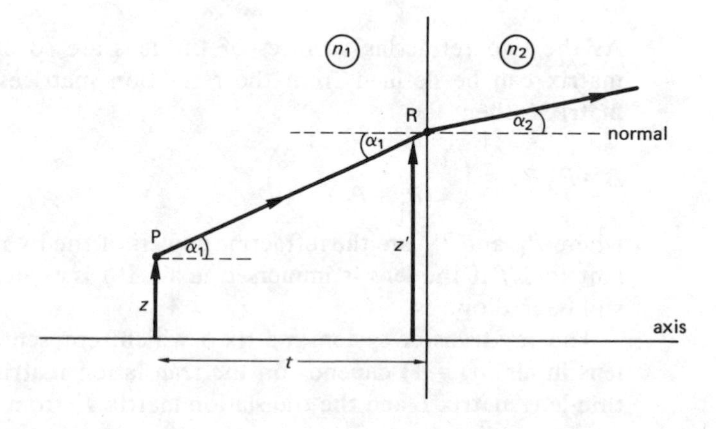

Figure 1.2

We obtain

$$z' = z + t\,\alpha_1$$
$$= z + (t/n_1)n_1\,\alpha_1$$

Also

$$n_2\,\alpha_2 = n_1\,\alpha_1$$

In matrix form, these two equations can be written

$$\begin{pmatrix} z' \\ n_2\alpha_2 \end{pmatrix} = \underbrace{\begin{pmatrix} 1 & t/n_1 \\ 0 & 1 \end{pmatrix}}_{\substack{\text{translation} \\ \text{matrix } T}} \begin{pmatrix} z \\ n_1\alpha_1 \end{pmatrix} \tag{1.7}$$

The determinant of T is 1.

If there are N different media of widths $t_1, \ldots, t_N$ and refractive indices $n_1, \ldots, n_N$, then the translation matrix for a ray passing right through them is

$$\begin{pmatrix} 1 & t_1/n_1 + \ldots + t_N/n_N \\ 0 & 1 \end{pmatrix}$$

or

$$\begin{pmatrix} 1 & \sum\limits_{m=1}^{N} T_m \\ 0 & 1 \end{pmatrix} \tag{1.8}$$

Similarly, the refraction matrix at a spherical surface of radius r can be shown to be

$$R = \begin{pmatrix} 1 & 0 \\ -(n_2 - n_1)/r & 1 \end{pmatrix} = \begin{pmatrix} 1 & 0 \\ -P & 1 \end{pmatrix} \tag{1.9}$$

(see, for example, *Introduction to Matrix Methods in Optics* by Gerrard and Burch).

P is the refracting power of the surface. It is measured in dioptres (D) if r is in metres.

(a) Thin-lens Matrix

As the two refracting surfaces of the lens are so close to one another, a single matrix can be defined from the refraction matrices R_1 and R_2. If we call this matrix L then

$$L = R_2 R_1 = \begin{pmatrix} 1 & 0 \\ -(P_1 + P_2) & 1 \end{pmatrix} \tag{1.10}$$

where P_1 and P_2 are the refracting powers of the two surfaces. $(P_1 + P_2)$ is equivalent to $1/f$ if the lens is immersed in air. If f is expressed in metres then $(P_1 + P_2)$ will be in dioptres.

The ray-transfer system matrix S which represents image formation by a thin lens in air ($n_A = 1$) depends on the translation matrix T_o from object to lens, the thin-lens matrix L and the translation matrix T_i from lens to image. Thus

$$S = T_i L T_o \tag{1.11}$$

(b) Generalised System Matrix

The procedure outlined in Section 1.4(a) can be extended to any optical system. Thus S is a generalised system matrix, given by

$$\begin{pmatrix} z' \\ n_2 \alpha_2 \end{pmatrix} = S \begin{pmatrix} z \\ n_1 \alpha_1 \end{pmatrix}$$

$$= \begin{pmatrix} A & B \\ C & D \end{pmatrix} \begin{pmatrix} z \\ n_1 \alpha_1 \end{pmatrix} \tag{1.12}$$

Here, n_1 is the refractive index of the medium on the object (input) side and n_2 is the refractive index of the medium on the image (output) side. In many problems both z and z' can be taken to be zero, the axial object and image points.

(i) *Properties of S*

1 $A = 0$

This means that

$$z' = B\,\alpha_1 \tag{1.13}$$

Its significance lies in the fact that all rays which are incident on the optical system at the *same* angle are focused to the *same* point, a distance z' above the axis. This occurs in the *secondary focal plane* of the system.

2 $B = 0$

Now

$$z' = A\,z \tag{1.14}$$

All rays leaving a common point in the object plane pass through the same point in the image plane; A is a measure of the *magnification* of the system. The matrix element B enables the image distance to be determined.

3 $C = 0$

This leads to

$$n_2\,\alpha_2 = D\,n_1\,\alpha_1 \tag{1.15}$$

All rays entering the system parallel to one another leave it as a parallel beam in a different direction. It is called the *telescopic* arrangement. The ratio: $n_2 \alpha_2 / n_1 \alpha_1$ is the angular magnification.

4 $D = 0$

This implies that

$$n_2\,\alpha_2 = C\,z \tag{1.16}$$

Now, all rays leaving a common point in the object plane leave the system as a parallel beam. Thus the object is in the *primary focal plane* of the system.

(c) Application to Thick Lenses and Lens Combinations

These optical systems can be analysed in a similar way to the thin-lens system if two reference planes RP1 and RP2 are set up near the first and last surfaces of the system. The system matrix is then defined *between* these planes, and can be expressed by

$$S = R_2 T R_1 \tag{1.17}$$

where R_1 and R_2 represent the respective refraction matrices at RP1 and RP2. Table 1.1 gives the results of a detailed analysis for a lens system in air.

Table 1.1

System parameter	*Measured* *from*	*to*	*Function of matrix element*
First focal point	RP1	F_1	D/C
First focal length	F_1	H_1	$-1/C$
First principal point	RP1	H_1	$(D-1)/C$
First nodal point	RP1	N_1	$(D-1)/C$
Second focal point	RP2	F_2	$-A/C$
Second focal length	H_2	F_2	$-1/C$
Second principal point	RP2	H_2	$(1-A)/C$
Second nodal point	RP2	N_2	$(1-A)/C$

N.B. Distances measured from the reference planes are positive if measured in the same direction as the incident light (i.e. from left to right).

1.5 Worked Examples

Example 1.1

Find an expression for the lateral displacement of a ray incident on a plane-parallel block of refractive index n and thickness t at an angle ϕ with respect to the normal. You may assume that the block is immersed in air.

A ray of light entering a transparent block at an angle of incidence ϕ will leave the block at the same angle because the medium is the same on both sides of the block. The emergent ray will, however, be displaced from the incident direction. Snell's law of refraction will need to be applied at one surface of the block.

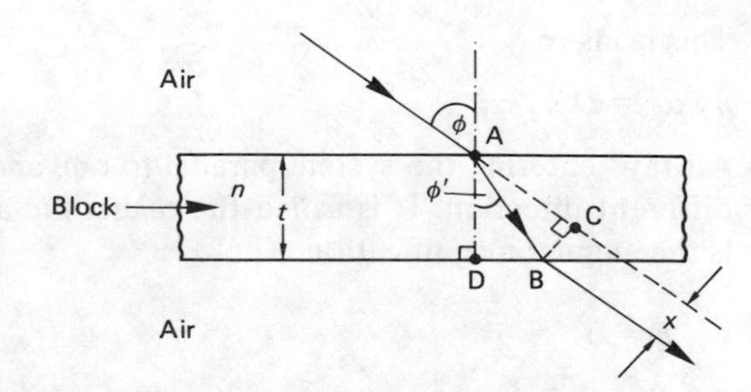

Data	*Given:*	ϕ
		n
		t
	Unknown:	x
Relevant equations		$n = \sin\phi/\sin\phi'$
		$x = \text{AB}\sin \text{B}\hat{\text{A}}\text{C}$ (1)

$\hat{BAC} = (\phi - \phi')$

To find AB, construct the right-angled triangle ADB, with angle ϕ' included, as in the diagram. Then

$$AB = t/\cos\phi' \tag{2}$$

and

$$x = (t/\cos\phi')\sin(\phi - \phi') \tag{3}$$

$$= (t/\cos\phi')[\sin\phi\cos\phi' - \cos\phi\sin\phi']$$

$$= t\sin\phi\,[1 - (\cos\phi/\cos\phi')(\sin\phi'/\sin\phi)]$$

$$= t\sin\phi\,[1 - \cos\phi/n\cos\phi']$$

$$= t\sin\phi\,[1 - \cos\phi/n\sqrt{1 - \sin^2\phi'}]$$

$$= t\sin\phi\,[1 - \cos\phi/\sqrt{n^2 - \sin^2\phi}]$$

NOTES

(i) Although the angle of refraction ϕ' had to be used in the solution it was necessary to remove it from the final expression because only ϕ was quoted in the question.

(ii) At normal incidence there is no refraction and x should be zero. This can be easily checked by putting $\phi = 0$, when $\sin\phi = 0$.

(iii) At grazing incidence $(\phi = \pi/2)$ and $x = t$.

(iv) As the final expression contains the refractive index n, this means that the lateral displacement will vary with wavelength λ because n depends on λ. In general the Cauchy relation

$$n = A + B/\lambda^2 + \ldots +$$

where A and B are constants characteristic of the block material, gives the dependence of n on λ in the normal dispersion region.

(v) If the block is immersed in a medium of refractive index n', other than air, then $\sin\phi/\sin\phi' = n/n'$.

Example 1.2

A ray of light is incident on a prism of refracting angle A and refractive index n at grazing incidence. It emerges from the prism at an angle θ to the surface. Prove that

$$n = (1 + 2\cos A\cos\theta + \cos^2\theta)^{\frac{1}{2}}/\sin A$$

The angle of refraction at the first surface of the prism is the critical angle, and the angle made with the normal on emergence at the second surface is $(90 - \theta)$. Snell's law will need to be applied at both surfaces. As the refractive index n is common to both expressions they can be equated to yield the required result.

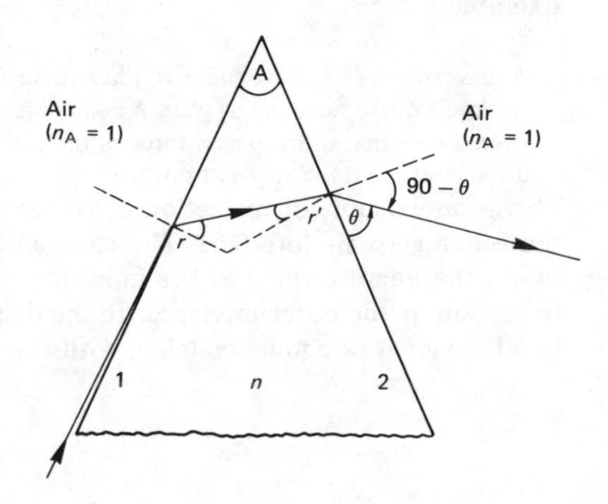

Data	*Given:*	angle of incidence = 90°
		angle of emergence = $(90 - \theta)$
		refracting angle = A
		angle of refraction at surface 1 = r
		angle of incidence at surface 2 = r'
	Unknown:	n
Relevant equations		$n = \sin i/\sin r = 1/\sin r = \cos\theta/\sin r'$
		$A = r + r'$

As

$$n = 1/\sin r = \cos\theta/\sin r' = \cos\theta/\sin(A - r)$$
$$n\sin(A - r) = \cos\theta$$

i.e.

$$n(\sin A\cos r - \cos A\sin r) = \cos\theta \quad (1)$$

Eliminate $\sin r$ and $\cos r$ using

$$\sin r = 1/n \text{ and } \cos r = (1 - \sin^2 r)^{\frac{1}{2}} = (1 - 1/n^2)^{\frac{1}{2}} = (n^2 - 1)^{\frac{1}{2}}/n$$

Then

$$n\{[(n^2 - 1)^{\frac{1}{2}}/n]\sin A - (1/n)\cos A\} = \cos\theta$$

and

$$(n^2 - 1)^{\frac{1}{2}}\sin A = \cos A + \cos\theta \quad (2)$$

from which

$$n^2 = (\sin^2 A + \cos^2 A + 2\cos A\cos\theta + \cos^2\theta)/\sin^2 A$$

and

$$n = (1 + 2\cos A\cos\theta + \cos^2\theta)^{\frac{1}{2}}/\sin A \quad (3)$$

NOTE

Here, again, as n is dependent on wavelength the angle θ will vary. The appearance of the spectrum of white light (blue light deviated more than red) is a consequence of this fact.

Example 1.3

A glass sphere (refractive index n_g) has an air-filled centre of radius a, which is equal to the thickness of the surrounding glass. A point object is situated on the inner wall and is observed from a direction which passes through the centre of the sphere. Prove that the image is at a distance $a(n_g - 1)/(3n_g - 1)$ from the object.

This is an example of refraction at two surfaces: the inner air/glass interface and the outer glass/air interface. The paraxial object–image relation must be applied twice; the image formed at the inner interface acts as an intermediate object for refraction at the outer interface. In the diagram below the light travels from right to left so great care must be taken with the algebraic signs.

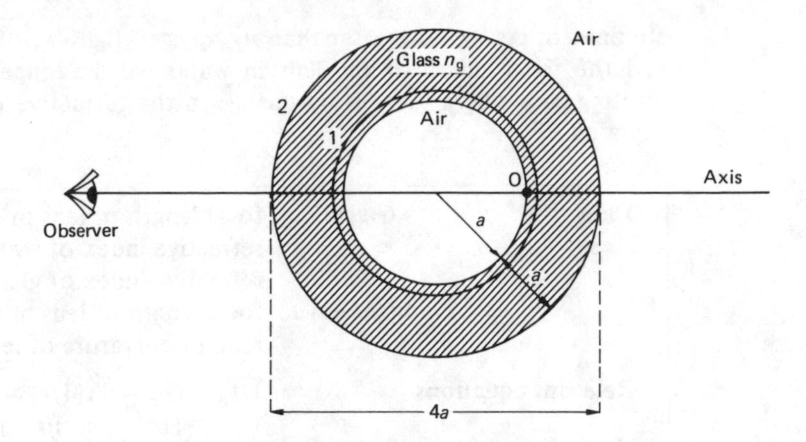

Data	*Given:*	refractive index of the glass n_g object distance from interface 1 = – 2*a* radius of curvature of interface 1 = + *a* radius of curvature of interface 2 = + 2*a*
	Unknown:	image position at interface 1 (≡ new object position at interface 2) final image position
Relevant equations		$1/x_o + n_g/x = (n_g - 1)/r_1$ at 1; $n_o/x_o + 1/x = (1 - n_g)/r_2$ at 2

Refraction at interface 1

$$-1/2a + n_g/x = (n_g - 1)/a$$

$$\therefore x = 2n_g a/(2n_g - 1) \qquad (1)$$

This is a positive quantity which means that the image is formed to the right of interface 1. This image now acts as the object for refraction at interface 2. The new object position x_o from interface 2 is

$$-(a + 2n_g a/(2n_g - 1)) = -(4an_g - a)/(2n_g - 1)$$

Refraction at interface 2

$$-n_g(2n_g - 1)/a(4n_g - 1) + 1/x = (1 - n_g)/2a$$

$$\therefore x = 2a(4n_g - 1)/(3n_g - 1) \qquad (2)$$

This is also a positive quantity so the image is formed to the right of interface 2. Therefore the distance between the object and the final image is

$$3a - 2a(4n_g - 1)/(3n_g - 1) = a(n_g - 1)/(3n_g - 1)$$

NOTE

You have to be especially careful in this example with the signs of the object and image distances because the light is travelling from right to left.

Example 1.4

A thin converging lens (n = 1.540) has a focal length of 40.0 cm in air (n_A = 1). What is the focal length of the lens when immersed in water (n = 1.330)?

The clue to the solution lies in the fact that the radii of curvature of the lens do not change when the lens is immersed in water. So applying the Lens Maker's equation to the lens in air and in water will allow the $(1/r_1 - 1/r_2)$ term to be eliminated. As n_w is greater than n_A rays of light will be refracted less severely and the focal length of the lens in water will be longer than that in air; in the limiting case where there is no change in the refractive index the rays will be undeviated.

Data	*Given:*	focal length of lens in air $f_A = 40.0$ cm
		refractive index of water $n_w = 1.330$
		refractive index of glass $n_g = 1.540$
	Unknown:	focal length of lens in water = f_w
		radii of curvature of lens = r_1, r_2
Relevant equations		$1/f_A = (n_g - 1)(1/r_1 - 1/r_2)$
		$1/f_w = [(n_g - n_w)/n_w]\ (1/r_1 - 1/r_2)$

We have

$$(1/f_A)(1/f_w) = (n_g - 1)\,n_w/(n_g - n_w) \tag{1}$$

so that

$$\begin{aligned} f_w &= f_A n_w (n_g - 1)/(n_g - n_w) \\ &= 40.0 \times 1.330 \times (1.540 - 1)/(1.540 - 1.330) \\ &= 139 \text{ cm} \end{aligned} \tag{2}$$

NOTE
The quoted refractive index n_w is characteristic of pure water. In the presence of impurities, dissolved ions, etc., as exist in tap water, the value may rise to 1.350 or 1.360.

Example 1.5

A biconvex lens has radii of curvature equal to 20.0 cm. The lens is fitted into the side of a water tank. If a point object is placed 100.0 cm from the lens along the axis, find the image position ($n_g = 1.500$, $n_w = 1.340$).

As the media on the two sides of the lens are different, this means that the generalised form of the paraxial lens equation must be used to determine the image position.

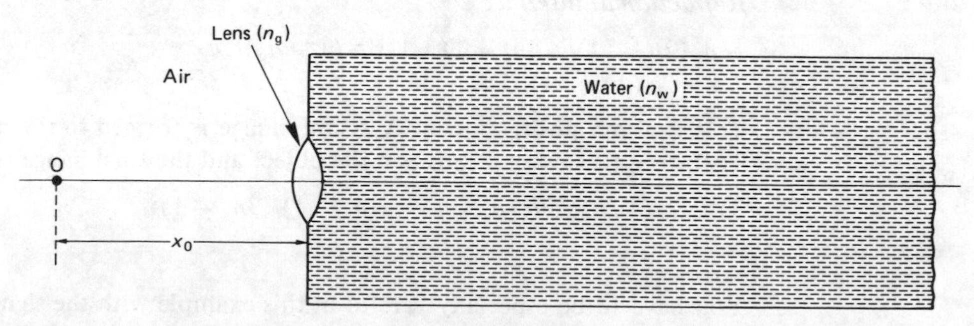

Relevant equation		$1/x_o + n_w/x = (n_g - 1)/r_1 + (n_w - n_g)/r_2$
Data	*Given:*	$r_1 = +\ 20.0$ cm
		$r_2 = -\ 20.0$ cm
		$x_o = 100.0$ cm
		$n_g = 1.500$
		$n_w = 1.340$
	Unknown:	x

We can express x as

$$n_w/x = (n_g - 1)/r_1 + (n_w - n_g)/r_2 = 1/x_o$$

which gives

$$x = n_w/[(n_g - 1)/r_1 + (n_w - n_g)/r_2 - 1/x_o] \quad (1)$$
$$= 1.340/[(0.500/20.0) + (-0.160/20.0) - 1/100]$$
$$= 134.0/[2.50 + 0.80 - 1]$$
$$= 58.3 \text{ cm}$$

Example 1.6

A converging glass lens (n_g = 1.500) has a focal length of 2.00 m when immersed in water (n_w = 1.330). What will the focal length of an 'air-bubble' lens be in water if it has the same radii of curvature and dimensions as the glass lens? Is it still converging?

A glass lens of given radii of curvature is to be replaced by a region in the water having the same radius of curvature but filled with air. The radii of curvature are the clue to the solution because as they are not quoted they must be eliminated during the analysis. The Lens Maker's equation is applied twice.

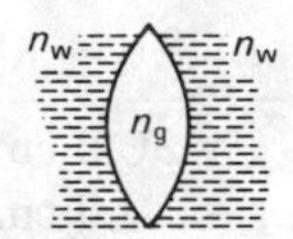

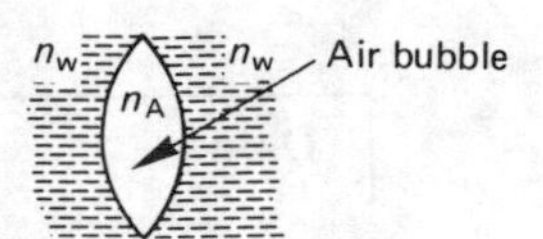

Data	*Given:*	$n_g = 1.500$
		$n_w = 1.330$
		$(f_g)_w = 2.00$ m
	Unknown:	$(f_A)_w$
		r_1
		r_2
Relevant equation		$1/f = ((n' - n'')/n')(1/r_1 - 1/r_2)$

For the glass lens in water:

$$1/f_g = ((n_g - n_w)/n_w)(1/r_1 - 1/r_2) \quad (1)$$

For the 'air-bubble' lens in water:

$$1/f_A = ((n_A - n_w)/n_w)(1/r_1 - 1/r_2) \quad (2)$$

so

$$(1/f_g)/(1/f_A) = (n_g - n_w)/(n_A - n_w)$$

and

$$f_A = f_g(n_g - n_w)/(n_A - n_w) \quad (3)$$
$$= 2.00 \times (1.500 - 1.330)/(1.0000 - 1.330)$$
$$= -2.00 \times 0.170/0.330$$
$$= -1.02 \text{ m}$$

The 'air-bubble' lens is diverging because in the algebraic sign convention being used concave lenses always have a negative focal length.

Example 1.7

An axial point source of light lies at a distance of 30.0 cm from a converging lens of focal length 20.0 cm. A large glass (n_g = 1.500) block has a plane face arranged perpendicular to the axis of the system and 10.0 cm behind the lens. Find the distance of the image from the lens.

In the absence of the glass block the image position can be found using the thin lens equation. However, when the block is present the image position lies further from the lens because the light leaving the lens is refracted towards the normal at the surface of the block. Snell's law of refraction will need to be applied at this surface even though we do not know the height of the ray above the axis.

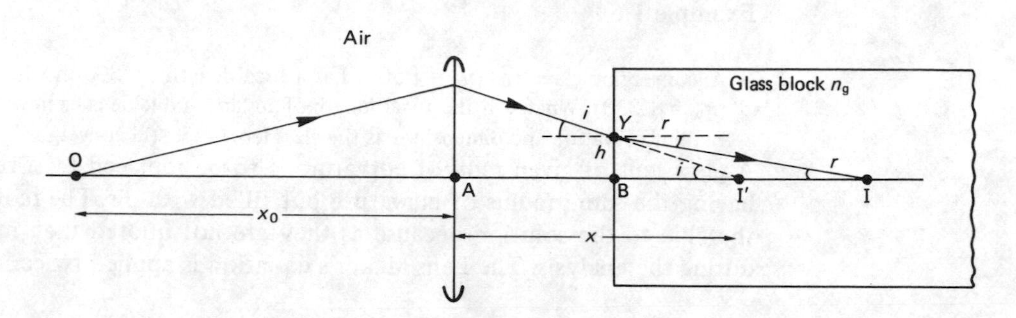

Data	*Given:*	x_o = 30.0 cm
		f = 20.0 cm
		AB = 10.0 cm
		n_g = 1.500
	Unknown:	x
		h
		AI
Relevant equations		$1/x_o + 1/x = 1/f$
		$n_g = i/r$

As

$$1/x = 1/f - 1/x_o$$

$$x = x_o f/(x_o - f) \quad (1)$$

and

$$\text{BI}' = [x_o f/(x_o - f) - \text{AB}] \quad (2)$$

Using trigonometry:

$$\begin{cases} \sin i \simeq \tan i \simeq i = h/\text{BI}' \\ \sin r \simeq \tan r \simeq r = h/\text{BI} \end{cases}$$

giving

$$n_g = i/r = \text{BI}/\text{BI}'$$

Therefore

$$\begin{aligned} \text{BI} &= n_g \times \text{BI}' \\ &= n_g\,[x_o f/(x_o - f) - \text{AB}] \quad (3) \\ &= 1.500 \times [(30.0 \times 20.0/10.0) - 10.0] \\ &= 75.0 \text{ cm} \end{aligned}$$

Hence AI = 85.0 cm

Example 1.8

A thin biconvex lens, constructed from crown glass, has radii of curvature of magnitude 20.0 cm. When the lens is in air an image for yellow light is formed 40.0 cm from the lens. Given that the refractive indices n_B and n_R for blue and red light are, respectively, 1.501 and 1.509, calculate the extent of the chromatic aberration for these wavelengths.

The refractive index of a medium is a function of wavelength – a fact that can be confirmed using a prism spectrometer. Thus each wavelength in the source will form an image at a different position; the image for blue light is closer to the lens than the image for red light. The linear separation of these images is a measure of the chromatic aberration for this range of wavelengths.

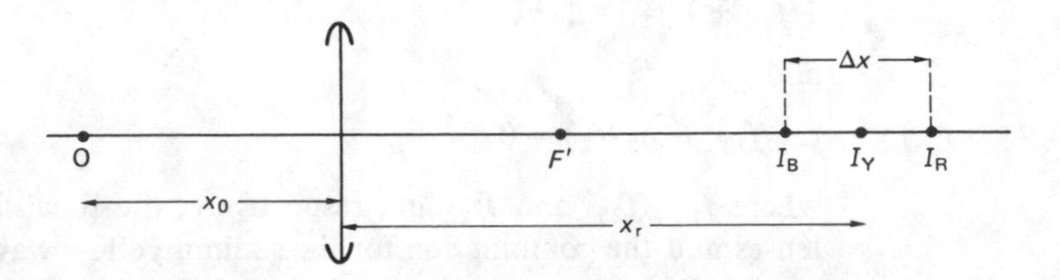

Relevant equation		$1/x_o + 1/x = 1/f = (n-1)(1/r_1 - 1/r_2)$
Data	*Given:*	$r_1 = +20.0$ cm
		$r_2 = -20.0$ cm
		$x_Y = 40.0$ cm
		$n_B = 1.501$, $n_R = 1.509$ } $\Delta n_{BR} = 0.008$
	Unknown:	x_o
		f
		n_Y, x_B, x_R } Δx

$$1/x_o + 1/x = (n-1)(1/r_1 - 1/r_2) \tag{1}$$

As x_o is fixed, differentiate this expression with respect to wavelength λ, when

$$-\left(\frac{1}{x^2}\right)\frac{dx}{d\lambda} = \left(\frac{1}{r_1} - \frac{1}{r_2}\right)\frac{dn}{d\lambda} \tag{2}$$

Omitting the negative sign and simplifying, we have

$$dx = x^2(1/r_1 - 1/r_2)\,dn \tag{3}$$

If x is identified with x_Y, then dx is equivalent to Δx_{BY}, the image separation for blue and yellow light. Then the difference between the refractive indices is Δn_{BY}. So (3) becomes

$$\Delta x_{BY} = x_Y^2(1/r_1 - 1/r_2)\,\Delta n_{BY} \tag{4}$$

Similarly, between yellow and red light the image separation is Δx_{YR} and the difference between the refractive indices is Δn_{YR}. Therefore the total linear separation of the images for blue and red light is

$$\begin{aligned}\Delta x &= \Delta x_{BY} + \Delta x_{YR}\\ &= x_Y^2(1/r_1 - 1/r_2)(\Delta n_{BY} + \Delta n_{YR})\\ &= x_Y^2(1/r_1 - 1/r_2)\,\Delta n_{BR} \qquad (5)\\ &= (40.0)^2 \times [1/20 - 1/(-20)] \times 8 \times 10^{-3}\ \text{cm}\\ &= 1.3\ \text{cm}\end{aligned}$$

NOTES

(i) The negative sign in the differentiated expression (2) means that since n decreases as wavelength increases x must increase with wavelength. The sign is omitted in (3) because we are only interested in the magnitude of the change in x.

(ii) A single thin lens will never be free from chromatic aberration because n varies with λ. It can be reduced, however, by placing a converging and a diverging lens in contact – the combination is known as an *achromat*. The focal lengths of the lenses can be calculated if the effective focal length of the achromat and the materials from which they are to be made are known. The relevant equations to be solved are:

$$1/f_{1Y} + 1/f_{2Y} = 1/F_Y$$

and

$$\omega_1/f_{1Y} + \omega_2/f_{2Y} = 0$$

where f_{1Y}, f_{2Y} and F_Y are, respectively, the focal lengths of the individual lenses and the combination for the sodium yellow wavelength and ω_1 and ω_2 are the dispersive powers of the lenses, defined by $(n_B - n_R)/(n_Y - 1)$.

Example 1.9

Find the positions of the principal planes of a thick spherical lens of radius R and refractive index n when immersed in air.

The two principal planes together with the focal and nodal planes make up the six cardinal planes of a thick lens. If their positions are known then the passage of a light ray through the lens can be specified completely. Indeed, the lens structure can be dispensed with. As the thick lens is in air, the nodal and principal planes coincide. In obtaining an expression for the position of each principal plane the latter are assumed to lie within the lens. However, an algebraic convention is necessary (see Section 1.3(b)). If the principal plane lies to the right of its lens vertex then the distance between them is said to be positive. In this example we might expect the principal planes to coincide and pass through the optical centre of the lens.

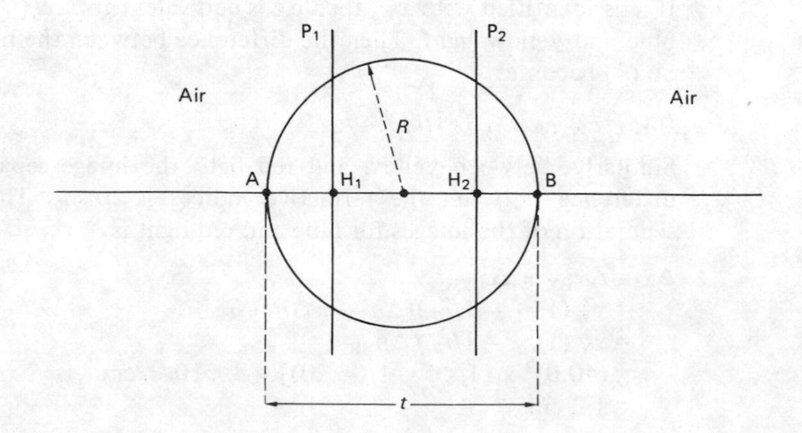

Data	*Given:*	$t = 2R$
		$r_1 = +R$
		$r_2 = -R$
	Unknown:	f (focal length of the thick lens)
		f_1' (secondary focal length of surface 1 – which is positive)
		f_2' (secondary focal length of surface 2 – which is negative)
		AH_1
		BH_2
Relevant equations		$1/f = (n-1)(1/r_1 - 1/r_2) + t(n-1)^2/nr_1r_2$
		$n'/x_o + n''/x = (n'' - n')/r$
		$AH_1 = -ft/f_2'$
		$BH_2 = -ft/f_1'$

$$1/f = (n-1)(1/R - 1/(-R)) + 2R(n-1)^2/nR(-R)$$
$$= 2(n-1)/nR$$
$$f = nR/2(n-1) \qquad (1)$$

f_1' can be found from

$$1/x_o + n/x = (n-1)/r_1 \qquad (2)$$

if $x_o = \infty$. Then

$$n/f_1' = (n-1)/R$$

and

$$f_1' = nR/(n-1) \qquad (3)$$

In a similar way we find that

$$f_2' = -nR/(n-1) \qquad (4)$$

Therefore

$$AH_1 = ft(n-1)/nR = nR \times 2R \times (n-1)/2(n-1) \times nR$$
$$= +R$$

and

$$BH_2 = -ft(n-1)/nR$$
$$= -R$$

The principal planes coincide at the centre of the thick lens, as we suspected.

NOTES

(i) The algebraic sign of the displacement of a principal plane from its particular lens vertex needs careful attention.

(ii) As the nodal planes coincide with the principal planes when the lens is in air, any ray directed towards the optical centre travels along a normal. The ray must leave along a normal also, in this case a continuation of the incident direction.

Example 1.10

A thin convex lens of focal length 5.0 cm is separated from a thin concave lens of focal length 10.0 cm by a distance of 15.0 cm in air. By treating this lens combination as a single thick lens, calculate the position of the cardinal planes.

The two-lens combination ought to be drawn roughly to scale. Then by taking an object at infinity the path of the ray through the lens can be traced. The position of the principal planes can be found by extrapolating the incident and emergent rays until they meet. Standard formulae may be used to obtain the displacements of the principal planes from their respective lenses if the focal length of the combination is known. Part of the problem is already answered because, as with Example 1.9, the lenses are in air, which means that the nodal planes occupy the same positions as the principal planes.

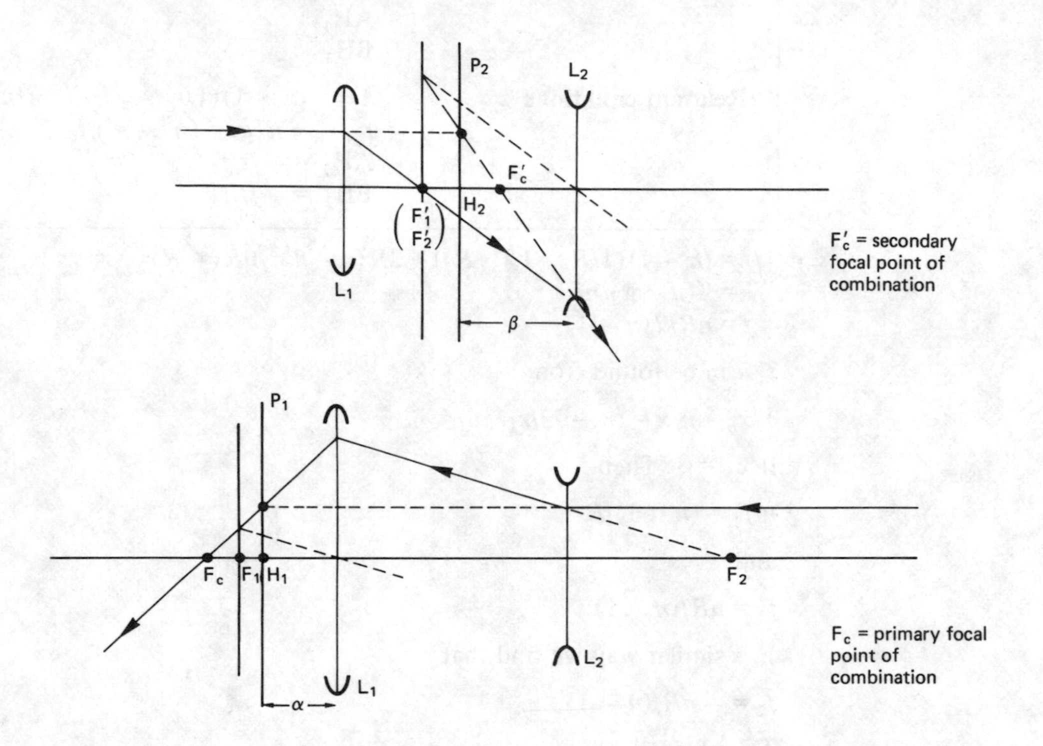

Data	*Given:*	$f_1 = +5.0$ cm
		$f_2 = -10.0$ cm
		$d = 15.0$ cm
	Unknown:	f_c
		α
		β
Relevant equations		$1/f_c = 1/f_1 + 1/f_2 - d/f_1 f_2$
		$\alpha = f_c d/f_2$
		$\beta = -f_c d/f_1$

$$1/f_c = (f_1 + f_2 - d)/f_1 f_2$$

and

$$f_c = f_1 f_2/(f_1 + f_2 - d) \tag{1}$$

which means that

$$\alpha = f_1 d/(f_1 + f_2 - d) \tag{2}$$

and

$$\beta = -f_2 d/(f_1 + f_2 - d) \tag{3}$$

Therefore,

$f_c = (+5.0) \times (-10.0)/(+5.0 - 10.0 - 15.0)$
$= 2.5$ cm
$\alpha = (+5.0) \times (15.0)/(-20.0)$
$= -3.8$ cm

and

$\beta = -(-10.0) \times (15.0)/(-20.0)$
$= -7.5$ cm

Thus P_1 lies 3.8 cm to the left of L_1 and P_2 lies 7.5 cm to the left of L_2. The distance between P_1 and P_2 is 11.3 cm.

NOTE
The calculated values of α and β enable the positions of P_1 and P_2 to be determined unequivocally. Inspection of the diagrams indicates that the algebraic signs of the values of α and β are correct.

Example 1.11

Using the 2-lens system of Example 1.10, determine where an object must be placed to give an overall magnification of +5.

Now that the principal planes have been found, the object distance can be measured from P_1 and the image distance from P_2. A lens equation similar in form to that for the thin lens may be used to solve the problem.

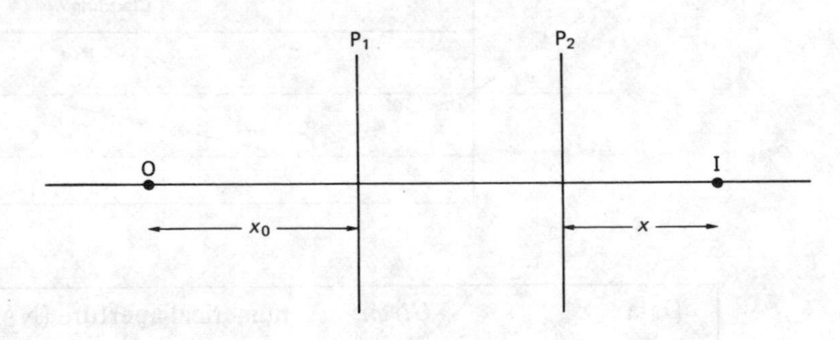

Data	*Given:*	$f_c = 2.5$ cm
		$m = +5$
	Unknown:	x_o
		x
Relevant equations		$1/x_o + 1/x = 1/f_c$
		$m = x/x_o$

Now

$$x = mx_o \tag{1}$$

and

$$1/x_o + 1/mx_o = 1/f_c$$

from which

$$x_o = (m+1)f_c/m \tag{2}$$

$= (5+1) \times 2.5/5$
$= 3.0$ cm

NOTE

If the image is inverted, i.e. $m = -5$, then a similar calculation shows that the object must be 2.0 cm from the first principal plane. Why not try this for yourself?

Example 1.12

Light is incident on a cylindrical step-index optical fibre, for which the difference in refractive index between the fibre and cladding is 1.00%. Calculate the refractive index of the fibre if its numerical aperture is 0.18 in air.

The light can enter the core of the fibre within a cone of half-angle θ_{max} and be propagated down it by a series of internal reflections at the core/cladding interface; θ_{max} defines the angle for which the light inside the fibre arrives at this interface at the critical angle. The light rays which we shall consider are referred to as the meridional rays as they pass through the axis of the fibre; other rays are known as skew rays. The numerical aperture is defined as $n_m \sin\theta_{max}$, where n_m is the refractive index of the medium outside the fibre. In air, n_m will be 1.0000 and the numerical aperture is simply equal to $\sin\theta_{max}$. It can be made larger by immersing the fibre in a medium of high refractive index. In the limit when $\sin\theta_{max}$ is 1.0000 the numerical aperture depends on the value of n_m.

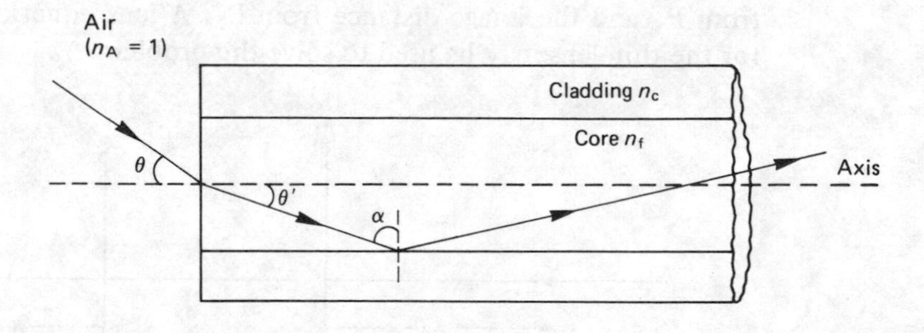

Data	*Given:*	numerical aperture (NA) = 0.18
		$n_f - n_c = 0.0100$
		$n_A = 1$
	Unknown:	n_f
		n_c
		θ, θ', α
Relevant equations		$\text{NA} = n_m \sin\theta$ (in general)
		$\sin\theta/\sin\theta' = n_f$ at entrance to fibre
		$1/\sin\alpha_c = n_f/n_c$ at core/cladding interface when α is equal to its critical value α_c

At the entry to the fibre

$$\sin\theta = n_f \sin\theta' \tag{1}$$

Now, as θ increases to its maximum value, θ' also increases. This means that α decreases to its minimum value α_c. So we can put

$$\begin{aligned}\sin\theta'_{max} &= \sin(90 - \alpha_c)\\ &= \cos\alpha_c\end{aligned} \tag{2}$$

Using (1)

$$\sin\theta_{\max} = n_f \cos\alpha_c$$
$$= n_f (1 - \sin^2\alpha_c)^{\frac{1}{2}}$$
$$= n_f (1 - n_c^2/n_f^2)^{\frac{1}{2}}$$
$$= (n_f^2 - n_c^2)^{\frac{1}{2}} \qquad (3)$$

If the difference between n_c and n_f is very small, as it is here, then we can write

$$\sin\theta_{\max} = [(n_f + n_c)(n_f - n_c)]^{\frac{1}{2}}$$
$$= (2n_f\,\Delta n_f)^{\frac{1}{2}} \qquad (4)$$

where Δn_f is the difference between the refractive indices of the core and cladding.

Substituting, we have

$$0.18 = (2n_f \times 0.010)^{\frac{1}{2}}$$

Therefore

$$n_f = (0.18)^2/0.020$$
$$= 1.60$$

NOTES

(i) In the step-index fibre there is a sharp change in the refractive index at the core/cladding interface. There is another kind of fibre called the gradient refractive index (GRIN) fibre in which n_f decreases radially from the central axis.

(ii) A bundle of fibres is called a fibrescope. It can be bent into an arc, or twisted, in order that inaccessible parts of an object may be examined. Usually, the outer fibres in the bundle conduct light to the object whilst the more central fibres conduct information about the object to the observer. The degree of bending of the fibre is fairly critical because light may escape from individual fibres and 'cross-talk' occurs.

(iii) As the largest value of $\sin\theta_{\max}$ is 1.000 0, the maximum value of the numerical aperture is dependent on the refractive index of the medium in which the fibre is immersed. The numerical aperture gives a measure of the light-gathering ability of the fibre.

(iv) The analysis presented here is fairly crude because geometrical optics is only an approximation to wave optics. For a rigorous treatment of the behaviour of light in a fibre it is necessary to invoke the wave properties of light. This can be quite difficult.

Example 1.13

The equation representing the path of a light ray in a non-homogeneous medium is

$$x = A\sin(y/B)$$

Determine an expression for the refractive index n if it is assumed that n depends on x alone and has the value n_o at the plane $x = 0$.

When a light ray meets the interface between two media having distinct refractive indices it is refracted. Such is the case at an air/glass interface. In examples of this kind it is best to set up a Cartesian coordinate system (X-Y), where the X-axis points into the medium. Then, as the refractive index is varying, the medium can be divided into a series of narrow segments, parallel to the Y-axis (as in the diagram below), and within which the refractive index is constant. As the refrac-

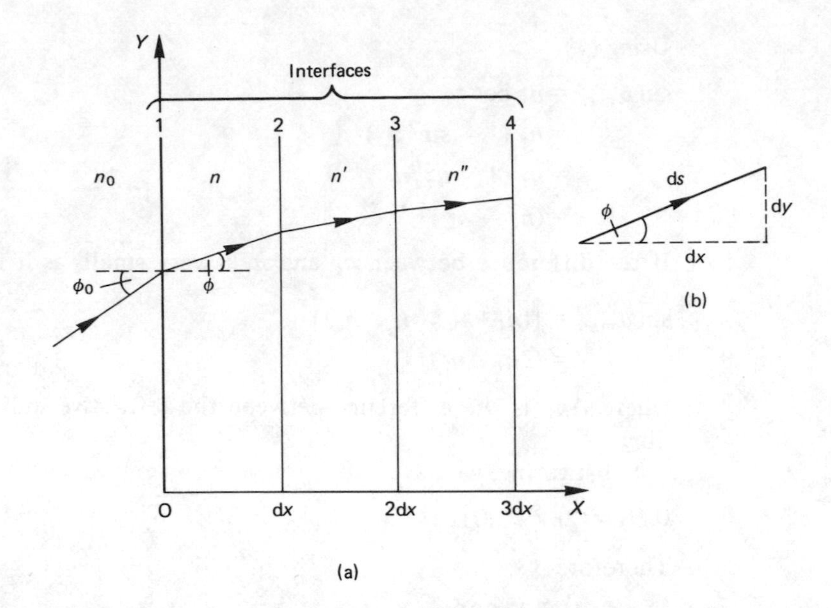

(a)

tive index is known exactly in one segment, the equation of the light ray can be established by applying Snell's law of refraction and the equation representing the length of a small element of the ray.

Data	*Given:*	$x = A \sin(y/B)$
		$n = n_o$ at $x = 0$
	Unknown:	ϕ_o
		ϕ
		equation for n

Let ds be the length of a small element of the ray. Then by Pythagoras' theorem:

$$ds^2 = dx^2 + dy^2 \quad (1)$$

or

$$(ds/dy)^2 = (dx/dy)^2 + 1$$

Applying Snell's law of refraction at interface 1, we have

$$n_o \sin \phi_o = n \sin \phi \quad (2)$$

As n is the variable, put

$$\sin \phi = dy/ds$$

then

$$1/\sin \phi = n/n_o \sin \phi_o = [(dx/dy)^2 + 1]^{\frac{1}{2}} \quad (3)$$

Transposing, (3) gives

$$(dx/dy)^2 = (n^2/n_o^2 \sin^2 \phi_o) - 1 \quad (4)$$

Now differentiate (4) with respect to y to give

$$2(d^2x/dy^2)(dx/dy) = (1/n_o^2 \sin^2 \phi_o)d(n^2)/dy$$
$$= (1/n_o^2 \sin^2 \phi_o)(d(n^2)/dx)(dx/dy)$$

or

$$d^2x/dy^2 = (1/2n_o^2 \sin^2 \phi_o)d(n^2)/dx \quad (5)$$

As

$$x = A \sin(y/B) \tag{6}$$

$$dx/dy = (A/B)\cos(y/B)$$

and

$$d^2x/dy^2 = -(A/B^2)\sin(y/B) = -x/B^2$$

Therefore, substituting into (5) gives

$$-x/B^2 = (1/2n_o^2 \sin^2 \phi_o)\, d(n^2)/dx$$

that is

$$d(n^2) = -(2xn_o^2 \sin^2 \phi_o/B^2)\, dx$$

Integration leads to the equation

$$n^2 = -x^2 n_o^2 \sin^2 \phi_o/B^2 + K \tag{7}$$

where K is a constant of integration. K can be evaluated by using the boundary condition: $n = n_o$ at $x = 0$ when

$$K = n_o^2$$

and (7) becomes

$$\begin{aligned} n^2 &= -x^2 n_o^2 \sin^2 \phi_o/B^2 + n_o^2 \\ &= -(n_o^2 A^2 \sin^2 \phi_o/B^2)\sin^2(y/B) + n_o^2 \end{aligned} \tag{8}$$

We now need to find ϕ_o. This can be done by using (4) and (6). Hence

$$\begin{aligned} (dx/dy)^2 &= (A^2/B^2)\cos^2(y/B) \\ &= (n^2/n_o^2 \sin^2 \phi_o) - 1 \end{aligned}$$

from which

$$1/n_o^2 \sin^2 \phi_o = (1/n^2)[1 + (A^2/B^2)\cos^2(y/B)] \tag{9}$$

Substitution into (8) will enable n to be determined as a function of y, i.e.

$$n^2 = \frac{-n^2 A^2 \sin^2(y/B)}{B^2[1 + (A^2/B^2)\cos^2(y/B)]} + n_o^2$$

On transposing terms, we arrive at

$$n^2(A^2 + B^2) = n_o^2 B^2 [1 + (A^2/B^2)\cos^2(y/B)]$$

and

$$n = [n_o B^2/(A^2 + B^2)^{\frac{1}{2}}]\,[1 + (A^2/B^2)\cos^2(y/B)]^{\frac{1}{2}} \tag{10}$$

Example 1.14

To repeat Example 1.5 using matrices.

There are three separate matrices involved here: a translation matrix from object to lens, the thin-lens refraction matrix, and a translation matrix from lens to image. The system matrix may then be obtained, and B (see Section 1.4(b)(i)) put equal to zero. The diagram is given in Example 1.5.

Data	As in Example 1.5 except that all distances and radii of curvature are expressed in metres.

The translation matrix T_o from object to lens is

$$T_o = \begin{pmatrix} 1 & x_o \\ 0 & 1 \end{pmatrix} \tag{1}$$

The thin-lens refraction matrix L is

$$L = \begin{pmatrix} 1 & 0 \\ -(P_1 + P_2) & 1 \end{pmatrix} \tag{2}$$

The translation matrix T_i from lens to image is

$$T_i = \begin{pmatrix} 1 & x/n_w \\ 0 & 1 \end{pmatrix} \tag{3}$$

Therefore the system matrix S is

$$S = \begin{pmatrix} 1 - (x/n_w)(P_1 + P_2) & (x/n_w)[1 - (P_1 + P_2)x_o] + x_o \\ -(P_1 + P_2) & 1 - x_o(P_1 + P_2) \end{pmatrix} \tag{4}$$

The image position will be found by putting $B = 0$, i.e.

$$(x/n_w)[1 - (P_1 + P_2)x_o] + x_o = 0$$

and

$$x = -n_w x_o/[1 - (P_1 + P_2)x_o] \tag{5}$$

$$P_1 = (n_g - 1)/r_1 = 0.500/0.20 = 2.50 \text{ D}$$
$$P_2 = (n_w - n_g)/r_2 = -0.160/-0.20 = 0.80 \text{ D}$$

Therefore

$$\begin{aligned} x &= -1.340 \times 1.00/(1 - 3.30) \\ &= -1.340/-2.30 \\ &= 0.58 \text{ m} \end{aligned}$$

NOTES

(i) $B = 0$ also gives us the magnification of the system, via

$$h' = Ah$$

Thus

$$\begin{aligned} A &= 1 - (x/n_w)(P_1 + P_2) \\ &= 1 - (0.58/1.340)(3.30) \\ &= -0.43 \end{aligned}$$

The negative sign implies that the image is inverted.

(ii) In a similar way the primary and secondary focal lengths can be found using the other rules in Section 1.4(b)(i).

Example 1.15

To repeat Example 1.7.

There are five individual matrices composing the system matrix S: a translation matrix T_o from object to lens, the thin-lens refraction matrix L, a translation matrix T_i (which would be sufficient to determine the position of the intermediate image), a refraction matrix R_b for the plane surface of the block and a translation matrix T_i from block to image.

Data As in Example 1.7 but with distances expressed in metres.

Now

$$\begin{pmatrix} z' \\ n_g \alpha_2 \end{pmatrix} = S \begin{pmatrix} z \\ \alpha_1 \end{pmatrix}$$

where

$$S = T_b\ R_b\ T_i\ L\ T_o \tag{1}$$

Therefore writing S in full we have:

$$S = \begin{pmatrix} 1 - (P_1 + P_2)\,AB - (x/n_g)(P_1 + P_2) & x_o + (AB + x/n_g)\,[1 - (P_1 + P_2)x_o] \\ -(P_1 + P_2) & 1 - (P_1 + P_2)\,x_o \end{pmatrix} \tag{2}$$

Putting element $B = 0$ we arrive at

$$x = n_g\,[-x_o/[1 - (P_1 + P_2)\,x_o]] - AB$$

and since, for a thin lens in air,

$$P_1 + P_2 = 1/f \text{ (in metres)}$$

the final result is

$$\begin{aligned} x &= n_g\,[-x_o/(1 - x_o/f) - AB] \\ &= 1.500\,[-0.30/(1 - 0.30/0.20) - 0.10] \\ &= 0.75 \text{ m} \end{aligned} \tag{3}$$

Example 1.16

To repeat Example 1.10.

As this is a lens combination, we need to set up two reference planes RP_1 and RP_2 between which the system matrix can be determined. It is sensible to choose the location of the convex lens for RP_1 and the location of the concave lens for RP_2. Then the system matrix S is composed of two thin-lens matrices L_1 and L_2 and a translation matrix T – so

$$S = L_2\ T\ L_1$$

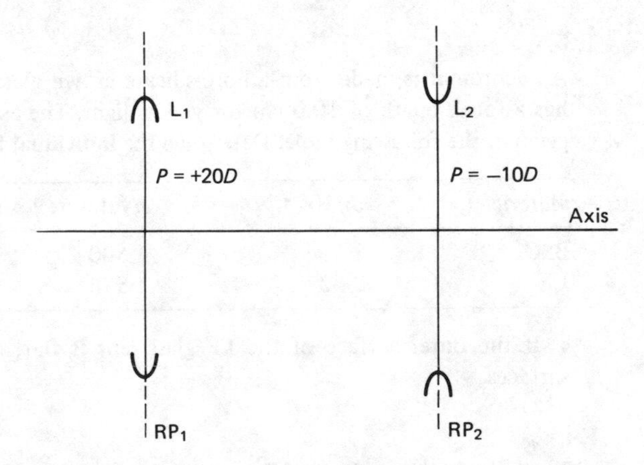

Data	See Example 1.10. The only difference here is that all distances are expressed in metres.

We shall substitute numbers into the matrices immediately because it makes the manipulation easier to carry out.

$$S = \begin{pmatrix} 1 & 0 \\ 10 & 1 \end{pmatrix} \begin{pmatrix} 1 & 0.15 \\ 0 & 1 \end{pmatrix} \begin{pmatrix} 1 & 0 \\ -20 & 1 \end{pmatrix}$$

$$= \begin{pmatrix} -2 & 0.15 \\ -40 & 2.5 \end{pmatrix}$$

Now we can determine the position of the cardinal points and planes using the rules in Table 1.1:

(a) The first focal plane is 1/40 m or 2.5 cm to the left of the first principal plane, and the second focal plane is 2.5 cm to the right of the second principal plane.
(b) The first principal plane is (2.5 − 1)/(− 40) m from the convex lens, i.e. 3.8 cm to its left.
(c) The second principal plane is (1 + 2)/(− 40) m from the concave lens, i.e. 7.5 cm to its left.

NOTE

It is important in calculations like this one to be aware of the sign convention. Look back at Sections 1.3(b) and 1.4(c) for clarification.

1.6 Questions

1.1

The left end of a long polymeric rod (n = 1.480) is ground and polished until a concave surface of radius 2.60 cm is formed. An object, 2.50 cm high, is placed at a distance of 12.0 cm from the vertex. Determine the primary and secondary focal lengths and the system magnification.

1.2

Use the data given in Example 1.3 but with the object situated at the diametrically opposite point on the inner wall. Prove that the object-image distance is still given by the same expression.

1.3

An achromat is made from a boro-silicate crown glass (BSC) and a light flint glass (LF) and has a focal length of 10.0 cm for yellow light. The essential characteristics of the glasses are given in the following table. Determine the individual focal lengths of the two lenses.

Material	$\omega/10^{-2}$	n (at λ 589.3 nm)
BSC	1.506	1.500
LF	2.427	1.576

If the outer surface of the LF glass lens is flat, calculate the radii of the three curved surfaces.

1.4

Treat the hollow sphere of Example 1.3 as a thick lens and obtain the image position. (*Hint:* Consider a small section of the sphere about the optical axis so that it looks like a negative-meniscus lens.)

1.5

Find the positions of the principal plane of a thick hemispherical lens of radius R and refractive index n immersed in air.

1.6

Find the cardinal planes of two converging lenses of focal lengths 5.00 cm and 10.00 cm separated by 10.00 cm in air.

1.7

The axial ray and an oblique ray characterised by an angle of incidence θ_{max} start to propagate down an optical fibre ($n_f = 1.500$) at the same instant. If T is the difference in time that they take to travel the same *axial* distance L, show that the multi-path time dispersion, denoted by T/L, of (a) an un-clad fibre is equal to 2.5 μs km^{-1}, and (b) a step-index fibre in which θ_{max} is 10.0° is 34.0 ns km^{-1}.

1.8

A light ray is incident on the interface between air ($n = 1.000$) and a variable refractive index medium at grazing incidence. Derive the equation of the ray in the medium if

$$n^2 = 1.000 + bx$$

1.9

A long glass rod of refractive index 1.500 has its left end ground to form a convex spherical surface of radius 1.00 cm. An object, 1.00 cm high, is 20.0 cm from this end of the rod in air. Determine the image position and the lateral magnification using the method of matrices.

1.10

An axial object point, a distance x_o from a thin converging lens forms an image at a distance x from the lens. If a sequence of object points forms a line of length L determine the longitudinal magnification using matrices.

1.11

Using the method of matrices obtain the position of the cardinal planes of a thick spherical lens of radius R and refractive index n in air. Compare your answers with those for Example 1.9.

1.12

The refracting angle of a prism is:
(a) the angle of refraction at the first surface of the prism;
(b) the angle between the sides of the prism which are concerned with the refraction process;
(c) obtained from Snell's law at grazing incidence;
(d) the sum of the angles of refraction at the two surfaces mentioned in (b);
(e) none of these but . . .

1.13

When white light is incident on a glass prism it:
(a) decomposes into the electromagnetic spectrum;
(b) passes through the prism undeviated;
(c) is dispersed into its component wavelengths with red light deviated more than blue;
(d) is dispersed into its component wavelengths with blue light deviated more than red;
(e) none of these but . . .

1.14

The refractive index of a glass prism is:
(a) related to wavelength by Cauchy's formula;
(b) the ratio of the speed of light in the prism to the speed of light *in vacuo*;
(c) a constant for all wavelengths;
(d) the ratio: sine of the angle of refraction to the angle of incidence for a given wavelength;
(e) none of these but . . .

1.15

The focal length of a thin lens is:
(a) a characteristic of the lens material;
(b) the same on both sides of the lens irrespective of the medium in which it is placed;
(c) dependent on the radii of curvature of the lens surfaces;
(d) dependent on the medium in which the lens is immersed as well as the geometry of its surfaces;
(e) none of these but . . .

1.16

Only a thick lens is subject to aberrations because:
(a) there is more material for the light to pass through;
(b) they are negligible in a thin lens;
(c) light rays never move in the paraxial regime;
(d) the dispersive powers of blue and red light are very large;
(e) none of these but . . .

1.17

The cardinal planes of a lens system are:
(a) only useful when rays move in the paraxial regime;
(b) extremely useful for analysing multiple lens systems;
(c) applicable to electrical and magnetic lenses also and enable the trajectories of electrons to be mapped out;
(d) a ray-tracing concept only and are not essential for understanding the operation of complicated lens systems;
(e) none of these but . . .

1.18

The propagation of light down an optical fibre is:
(a) a consequence of its gradient refractive index;
(b) due to the large numerical aperture;
(c) brought about completely by total internal reflection at the core-cladding interface;
(d) improved by increasing the numerical aperture;
(e) none of these but . . .

1.19

The reason that the sun can be seen after it has passed below the horizon is:
(a) because red light is refracted more than blue light thus giving the red colour to the sun;
(b) due to scattering of light by air molecules;
(c) because it is similar to the formation of a mirage in the desert;
(d) due to the light from the sun travelling through the lower denser regions of the atmosphere as quickly as possible and thus making the optical path length as small as possible; possible;
(e) none of these but . . .

1.20

When we are looking at an object with an optical system we must ensure that as much light enters the system as possible. This means that:
(a) the objective lens must be enormously large in diameter;
(b) the entrance pupil must be enormously large;
(c) the aperture stop must be as large as possible;
(d) the field stop must be as large as possible;
(e) none of these but . . .

1.21

The magnifying power of an optical instrument is defined as:
(a) the ratio of the size of the retinal image produced by the instrument to the size of the retinal image when the object is at the near point of the eye;

(b) the product of the lateral magnification of the objective lens and the angular magnification of the eyepiece;
(c) the ratio of the angle subtended at the eye by the final image to the angle subtended at the objective by the object;
(d) its ability to enlarge the object;
(e) none of these but . . .

1.7 Answers to Questions

1.1 – 5.42 cm; – 8.02 cm; 1/3 (inverted).
Use: $1/x_o + n'/x = (n' - n)/r = 1/f = n'/f'$
and $m = x/x_o$

1.3 (a) 3.80 cm; – 6.12 cm
Use the equation quoted in the Notes to Example 1.8 and obtain

$$f_{1Y} = F_Y(1 - \omega_1/\omega_2) \text{ and } f_{2Y} = -\omega_2 f_{1Y}/\omega_1$$

(b) $r_1 = 4.24$ cm; $r_2 = r_3 = -3.45$ cm
Apply the Lens Maker's equation to each lens.

1.4 Use the expression for the focal length of a thick lens given in Example 1.9 to obtain $f = -2n_g a/(n_g - 1)$. Also obtain $AH_1 = -a$ and $BH_2 = -2a$, so that both principal planes pass through the centre of the sphere.
The object distance relative to the first principal plane is a. Use

$$1/x_o + 1/x = 1/f$$

to give $x = 2n_g a/(1 - 3n_g)$. Hence the image–object distance may be determined.

1.5 $AH_1 = R/n$; $BH_2 = 0$

1.6 $f_c = +10.0$ cm; $\alpha = +10.0$ cm; $\beta = -20.0$ cm
This question is similar to Example 1.10.

1.7 Prove that $T/L = n_f(n_f - n_c)/cn_c$
In (a) put $n_c = 1$.
In (b) use $\sin\theta_{max} = 2n_f(n_f - n_c)^{\frac{1}{2}}$, as derived in Example 1.12.

1.8 Use a similar technique to Example 1.13. Take ϕ_o to be 90° and obtain

$$2x^{\frac{1}{2}} = b^{\frac{1}{2}}y + \text{constant } K$$

At $x = 0$, $y = 0$ and so $K = 0$. Hence obtain

$$x = by^2/4$$

1.9 3.3 cm; 0.1 (inverted).
The system matrix is

$$\begin{pmatrix} 1 - x/3 & 20 - 6x \\ -1/2 & -9 \end{pmatrix}$$

x is found by putting element B equal to zero. The lateral magnification is equal to element A.

1.10 The system matrix is $T_i \, L \, T_o$. Put element B equal to zero, and obtain the usual thin-lens equation

$$1/x_o + 1/x = 1/f$$

Differentiate implicitly to give

$$-\Delta x_o/x_o^2 - \Delta x/x^2 = 0$$

Letting Δx_o equal L, you obtain the longitudinal magnification as $\Delta x/L$, which is $(x/x_o)^2$. This is the square of the lateral magnification.

1.11 Let the reference planes RP_1 and RP_2 touch the poles of the thick lens. Then

$$S = R_2 \, T R_1$$

$$= \begin{pmatrix} 1 - 2(n-1)/n & 2R/n \\ -2(n-1)/nR & 1 - 2(n-1)/n \end{pmatrix}$$

Now use the rules listed in Table 1.1 of Section 1.4(b). For example the first principal plane is at a distance of $(D - 1)/C$ from reference plane RP_1. That is, $+R$, which agrees with the result given in Example 1.9.

1.12 (b)

1.13 (d)

1.14 (a)

1.15 (e)
(d) is not quite correct. The lens material is also important.

1.16 (e)
The image formed by a thin lens also suffers from aberrations.

1.17 (b); (c)

1.18 (c)

1.19 (d)

1.20 (c)

1.21 (a); (b); (c)
All of these are acceptable definitions: (a) is used with a simple magnifier; (b) with a compound microscope; (c) with an astronomical telescope.

2 Wave Motion

2.1 Wave Description of Light

The results of interference, diffraction and polarisation experiments carried out with wavelengths within and outside the visible spectrum can be adequately explained using a wave interpretation. In this description light has a transverse waveform in which the vibrating quantity is at right angles to the direction of propagation.

A reasonable representation of an unpolarised light wave can be obtained using a rope. Fix one end of the rope and 'flick' the other in one direction – a wave pulse will pass down the rope. Now vary the direction of 'flick' continuously in a completely random manner – a series of wave pulses will travel down the rope with varying planes of polarisation. To simulate light, this process would need to be carried out very rapidly – in times less than 10 ns or so. Then, a series of wave trains will be generated, polarised in planes chosen at random and with phases changing discontinuously in going from one wave train to the next.

2.2 Progressive and Stationary Waves

As the name suggests a progressive wave is not confined to a particular region of space but propagates throughout space, transporting energy with it. If the amplitude stays constant then it is called a plane-progressive wave whereas if it varies as $(\text{distance})^{\frac{1}{2}}$ it is a cylindrical wave and if as $(\text{distance})^{-1}$ a spherical wave (although the latter should be interpreted as a geometrical result based on the fact that the surface area of a sphere is $4\pi r^2$). Stationary waves are confined to a particular region of space. They can be produced by two progressive waves of equal frequencies moving in different directions, such as occurs in the particular case of a wave incident normally on a plane mirror. Nodes and anti-nodes will be established at fixed positions although the amplitude will vary with time – it takes its maximum value at times equal to $mT/2$, where m is 0 or an integer and T is the periodic time.

2.3 Wave Equations

All wave motions: electromagnetic, acoustic, seismic, etc., can be described by a second order differential equation of the kind:

$$\mathrm{d}^2y/\mathrm{d}x^2 = (1/v^2)\,\mathrm{d}^2y/\mathrm{d}t^2 \tag{2.1}$$

where y is the displacement, x is the distance, v is the speed of propagation and t is the time. This equation can be rather difficult to handle and so it is usual to take particular solutions as the starting point in any discussion of light waves.

$y = f(x + vt)$

can be shown to satisfy the general wave equation; the plus and minus signs refer to the direction of propagation: if it is along the positive x-axis the negative sign is taken, and vice versa.

A plane-progressive wave can be represented by

$$y = A \sin [(2\pi/\lambda)(x - vt) + \phi] \tag{2.2}$$

and the displacement y can be determined at any position x and at any time t. ϕ is the initial phase or epoch of the wave motion.

If the wave is frozen at some instant of time t, such as would be the case if a photograph were taken, and if X–Y axes are set up then the displacement at *any* position relative to the origin may be expressed by the relation

$$y = A \sin (2\pi x/\lambda + \phi') \tag{2.3}$$

where ϕ' is the value of the initial phase at the origin.

Alternatively, if the displacement is determined as a function of time at a *fixed* position x_f then it satisfies the relation

$$y = A \sin (2\pi/T + \phi'') \tag{2.4}$$

where ϕ'' is the initial phase at x_f at the moment the timings begin.

2.4 Principle of Superposition

This principle states that the resultant displacement of any point is the sum of the displacements of each individual wave. Although the resultant displacement can be found mathematically, it is easier, and perhaps more instructive, to use the graphical phasor technique. In this, a circle is drawn with a radius equal to the amplitude of the wave. Cartesian coordinate axes Y–Z are then set up, as shown in Fig. 2.1, with the positive X-axis into the plane of the paper.

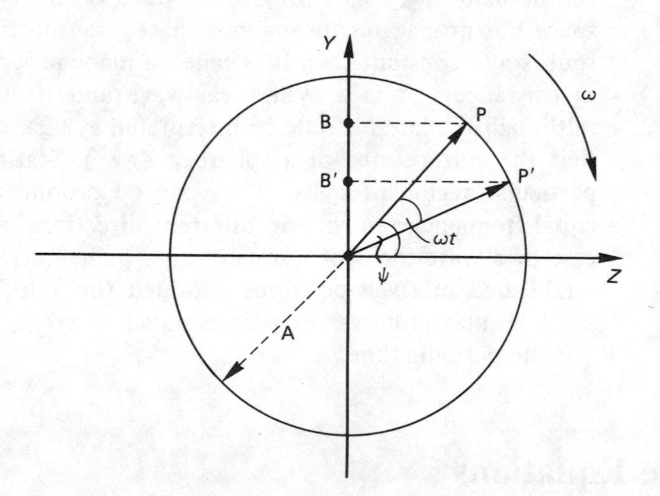

Figure 2.1

Figure 2.1 combines the two ways of looking at a wave: sideways-on and end-on, and is the basis for understanding simple harmonic motion. As point P moves around the circumference, the radius vector moves through an angle of 2π – behaving like the phase of the wave. The projection of P on the Y-axis, viz. point B, varies its distance from the Z-axis and takes all values between 0 and A – thus behaving like the displacement y. After a time t, point P moves to P′ and the radius vector makes an angle $(\psi - \omega t)$ with the Z-axis. Then

$$\text{OB}' = y = A \sin (\psi - \omega t) \tag{2.5}$$

Figure 2.1 is called a *phase-amplitude* or *phasor* diagram.

The position of point P can be determined by comparing (2.5) with (2.2), when

$$\psi = 2\pi x/\lambda + \phi \tag{2.6}$$

Two wave motions of the *same* frequency can be superposed if the phase difference δ between them is constant at all times. Then the sources producing them are referred to as *coherent*. Two circles are needed if the amplitudes are different: point P can be determined for each wave motion and the resultant amplitude found using either the 'parallelogram of forces' concept or the cosine rule, as in Fig. 2.2. Just as OP_1 and OP_2 move round the circumference of their respective circles with the same angular velocity ω, OP_R does also. Note that OP_1 and OP_2 keep the same angle δ between them at all times.

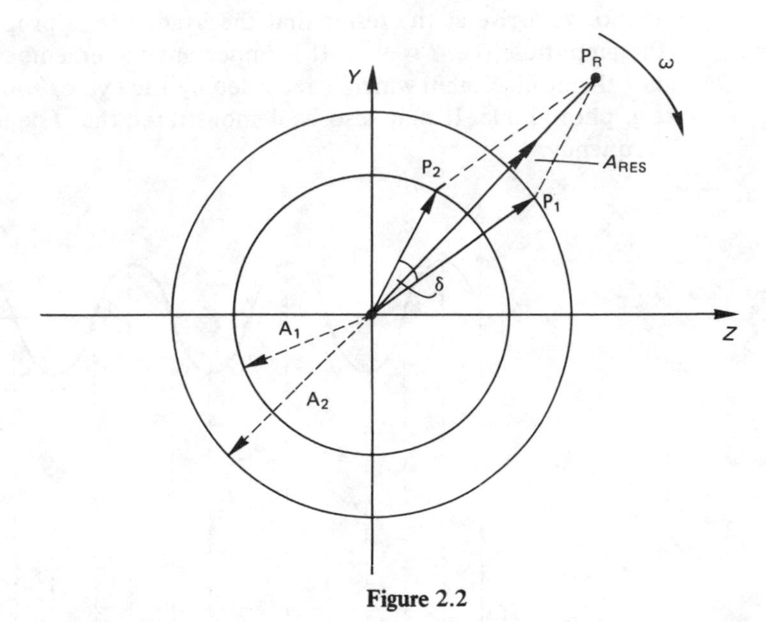

Figure 2.2

2.5 Optical Path Length

This is generally defined as the product of the refractive index n of the medium and the geometrical path length L travelled by the light. The speed of the waves in the medium is not equal to its value *in vacuo* which means that the wavelength λ_{med} is n^{-1} times its value λ *in vacuo*. As we shall see in the examples in Chapter 3, the optical path difference is of paramount importance in interference work. It must be used in preference to the geometrical path difference because although one wave may travel entirely through air a second wave may travel through several different media before finally interfering with the first wave. There can be complications. For example, a light wave may suffer a phase change of π on reflection at a dielectric surface during its path through the medium. Then

$$\text{optical path length} = (n \times \text{geometrical path length}) + \lambda/2n \tag{2.7}$$

2.6 Irradiance

As we have previously mentioned, progressive waves transport energy throughout space. The amount of energy flowing perpendicularly through unit area per unit time is called the irradiance, given the symbol I. The units are watts per square metre. In some textbooks the irradiance is called the intensity or areance.

A rigorous definition of irradiance concerns the value of y^2 averaged over an interval of time which is long compared with the period of the sinusoidal waveform. At a given position in space let us define y by

$$y = A \sin(\omega t + \phi)$$

Thus the irradiance I may be expressed as

$$\begin{aligned} I &\propto \langle y^2 \rangle \\ &\propto A^2 \langle \sin^2(\omega t + \phi) \rangle \end{aligned} \qquad (2.8)$$

where the brackets stand for time-averaging. Figure 2.3(a) shows the variation of y with time t and Fig. 2.3(b) illustrates the corresponding y^2 variation. The time-average value of y^2 is seen to be $A^2/2$; y^2 lies as much above this value as below it. So we arrive at the result that the irradiance is proportional to the square of the amplitude (i.e. $I \propto A^2$). It is important to remember that it is the irradiance not the displacement which is recorded by the eye or some other kind of detector, e.g. photodiode. It may also be demonstrated that I depends on the square of the frequency.

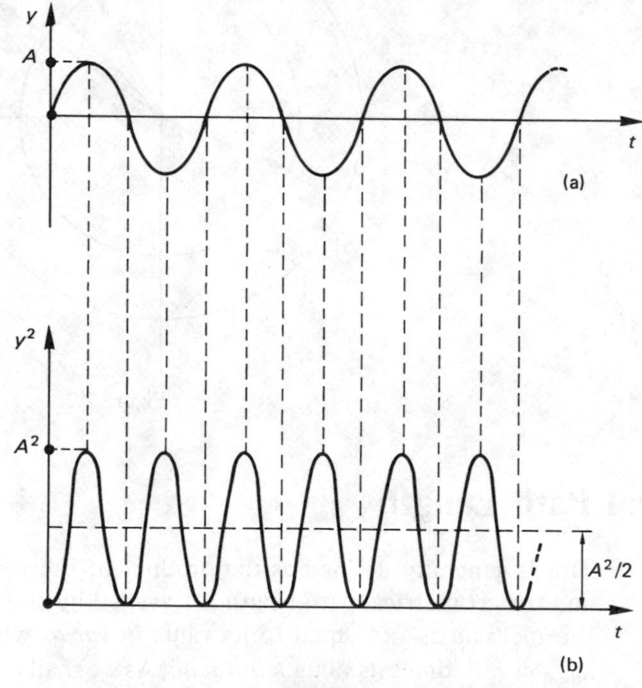

Figure 2.3

In order to remove the constant of proportionality between irradiance and amplitude^2, you will often see the term *normalised irradiance* used. This is just the irradiance determined at some particular point compared with the value obtained at some central, or reference, point. The normalised irradiance is, therefore, a ratio which is a pure number.

2.7 The Exponential Representation

The phasor diagram, discussed in Section 2.4, is an extremely useful aid for adding wave motions pictorially. It gives highly relevant information on the effect of two or more wave motions at some external point. Indeed we shall use this technique in Chapter 4 with the diffraction of light. There is another way of representing wave motions which is extremely beneficial to use when the sine and cosine formulations look more awkward in appearance and present difficulties in their manipulation. This is the exponential or complex representation. This representation is used widely in the electromagnetic theory of radiation, in quantum mechanics and at a more elementary level in alternating current theory.

A complex number P can be written as

$$P = Q + \mathrm{i}S \tag{2.8}$$

where Q and S are real numbers and i is an operator. P can be shown graphically using the Argand diagram shown in Fig. 2.4. It has a real axis and an imaginary axis. To locate P measure a distance Q along the real axis, then the operator i tells us to turn anti-clockwise through 90° and measure a distance S along the direction of the imaginary axis.

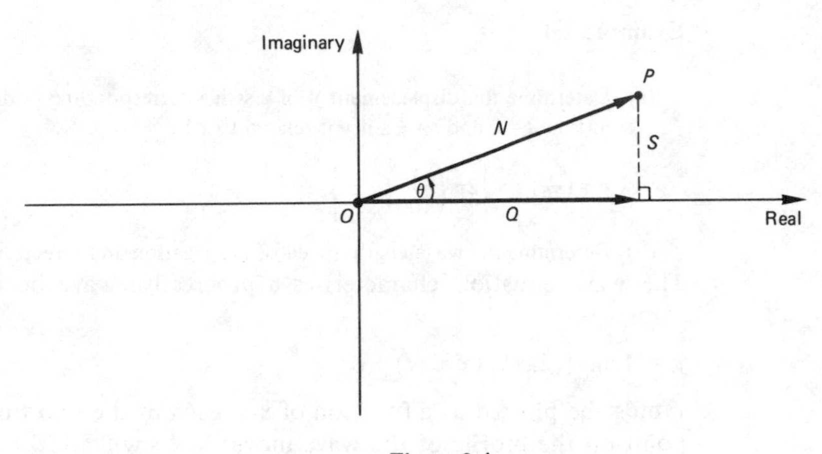

Figure 2.4

Let the vector OP have a length N and lie at an angle θ from the real axis, then

$$Q = N\cos\theta \quad \text{and} \quad S = N\sin\theta$$

so that (2.8) can be written

$$P = N(\cos\theta + \mathrm{i}\sin\theta) \tag{2.9}$$

Using de Moivre's theorem, (2.8) may be written

$$P = N\exp(\mathrm{i}\theta) \tag{2.10}$$

N is called the *modulus* or *magnitude* of P, and is equal to $(Q^2 + S^2)^{\frac{1}{2}}$. θ is the argument or phase angle of P and is equal to $\tan^{-1}(S/Q)$.

Suppose that θ is some function of time. Then this means that the position of P in the Argand diagram will vary. If, for example, θ varies with time in a sinusoidal way then P will rotate at a constant angular frequency ω whilst keeping the same distance from the origin. There is now a strong similarity with Fig. 2.1; OP is called a *phasor* or *rotating vector*.

We may write (2.5) as

$$\mathcal{Y} = A\exp[\mathrm{i}(\psi - \omega t)] \tag{2.11}$$

where $\mathcal{Y}$ is the complex displacement and ψ is $2\pi x/\lambda + \phi$. Then y is the projection of the end point of the phasor on the imaginary axis. As time goes on the distance of the projection from the real axis will vary. This is exactly what is required for a sinusoidally varying wave motion. If the wave motion is written as a cosine term then it is the projection of the phasor on the real axis which is required.

The exponentials allow mathematical computations to be made easier, especially when there are a large number to add (see Example 2.6). The resultant displacement is the imaginary part of the resultant complex displacement. We shall write this as

$$y = \mathrm{Im}\, \mathcal{Y} \qquad (2.12)$$

The irradiance is a real quantity and can be found by multiplying $\mathcal{Y}$ by its complex conjugate $\mathcal{Y}^*$.

2.8 Worked Examples

Example 2.1

(a) Determine the displacement y of a string corresponding to different positions x along it at time $t = 0$ and $t = \frac{1}{8}$ s if y is related to x by

$$y = (\tfrac{1}{20}) \sin \left\{ 2\pi \left(\frac{x}{100} \right) + 2t \right\}$$

(b) Determine the wavelength, speed of propagation and direction of motion.

The wave equation characterises a progressive wave because it has the general form

$$y = A \sin [(2\pi/\lambda)(x \pm vt) + \phi]$$

y must be plotted as a function of x at each of the two times. The distance that a point on the profile of the wave moves in $\frac{1}{8}$ s will allow the speed of propagation to be found. The direction of motion may also be established, even though it can be found by inspection of the wave equation; the positive sign in the expression tells us that the wave is moving along the negative x direction.

Data		
	Given:	$\phi = 0$
		$A = \frac{1}{20}$ m
	Unknown:	y at $t = 0$ and $t = \frac{1}{8}$ s
		v
		λ
Relevant equations		$y = (\frac{1}{20}) \sin \left\{ 2\pi \left(\frac{x}{100} \right) + 2t \right\}$
		$v = f\lambda$

At $t = 0$:

$$y = (\tfrac{1}{20}) \sin (2\pi x/100)$$

At $t = \frac{1}{8}$ s:

$$y = (\tfrac{1}{20}) \sin \left\{ 2\pi \left(\frac{x}{100} + \frac{1}{4} \right) \right\}$$

Tabulate y as a function of x, from 0 to 100 m, say, for both times – as follows:

x/m	y/m	x/m	y/m
0	0 ($t = 0$)	0	$\frac{1}{20}$ ($t = \frac{1}{8}$ s)
25	$\frac{1}{20}$	25	0
50	0	50	$-\frac{1}{20}$
75	$-\frac{1}{20}$	75	0
100	0	100	$\frac{1}{20}$

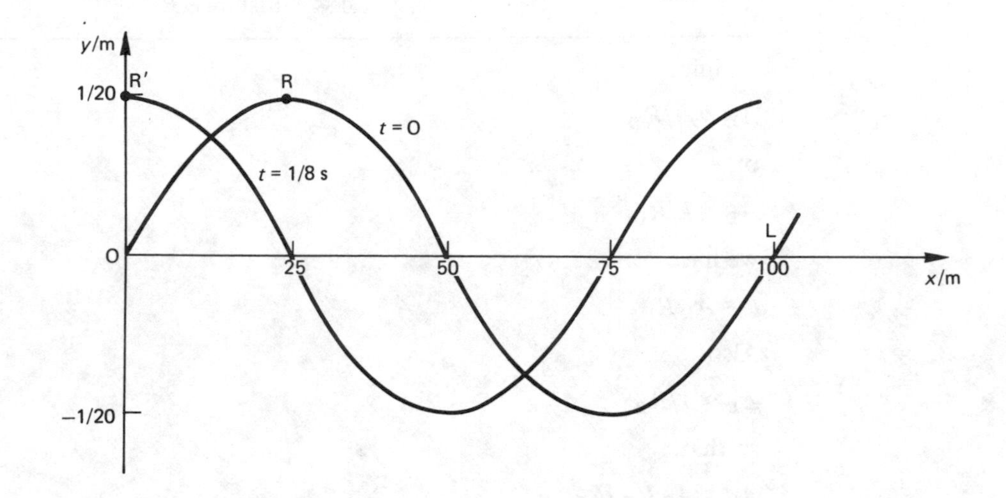

The wavelength is 0L = 100 m.
The wave moves a distance of 25 m to the left in $\frac{1}{8}$ s. For example, point R moves to R′. Therefore the speed of propagation of the wave is

$$v = 25/(\tfrac{1}{8}) \text{ m s}^{-1}$$
$$= 200 \text{ m s}^{-1}$$

Example 2.2

A point source emits waves and as they pass a point 100 cm from the source they generate a simple harmonic motion having the form

$$y = 0.36 \sin(2\pi t/T)$$

where T is $\frac{1}{20}$ s. If the waves travel at 80 cm s^{-1} find the equations of the waves 50 cm and 2.5 m from the source at time t.

The point source emits circular waves in two dimensions and spherical waves in three dimensions. As a consequence of the inverse square law (and surface area proportional to distance^2) the amplitude falls off inversely with distance from the source.

The equation of the wave is known 100 cm from the source at time t, so if we wish to know the equation 50 cm from the source at time t we must substitute a time $(t + 50/v)$ into the above expression. For this is the time that the wave arrives at the 100 cm point. With the 250 cm point we need to substitute a time of $(t - 150/v)$ into the above expression.

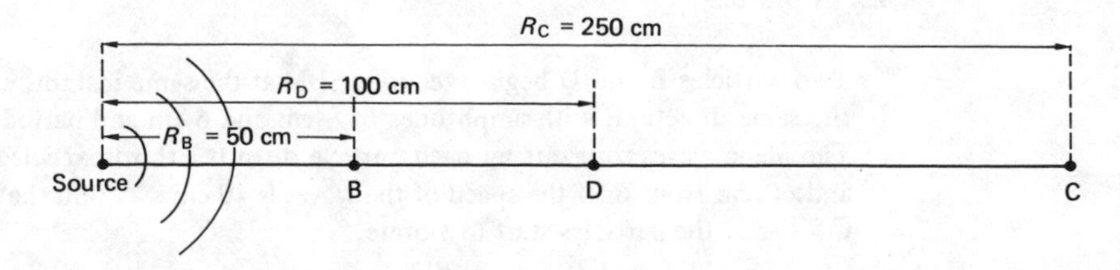

Data	*Given:*	speed of propagation $v = 80$ cm s^{-1} distances of points B, C and D from the source amplitude of wave at D = 0.36 cm period = $\frac{1}{20}$ s
	Unknown:	amplitudes at B and C – call these A_B and A_C transit times from B to D and from D to C.
Relevant equations		$y = 0.36 \sin (40\pi t)$ $A \propto 1/\text{distance } R$

Since

$$A_D \propto 1/R_D$$

or

$$A_D = k/R_D \quad (1)$$

we have

$$k = A_D R_D \quad (2)$$

Also

$$A_B \propto 1/R_B$$

so that

$$A_B = A_D R_D / R_B \quad (3)$$

and

$$A_C = A_D R_D / R_C \quad (4)$$

For the wave to be at B at time t it follows that it must reach D at the later time $(t + R_{BD}/v)$. Similarly, for the wave to be at C at time t it must leave D at the earlier time of $(t - R_{DC}/v)$. Therefore, the displacement y_B at B can now be determined from the displacement y_D at D. So

$$y_B = (A_D R_D / R_B) \sin [40\pi (t + R_{BD}/v)] \quad (5)$$

and

$$y_C = (A_D R_D / R_C) \sin [40\pi (t - R_{DC}/v)] \quad (6)$$

Now

$$A_B = 0.36 \times 100/50 = 0.72 \text{ cm};$$
$$A_C = 0.36 \times 100/250 = 0.14 \text{ cm};$$
$$R_{BD}/v = \tfrac{50}{80} = \tfrac{5}{8} \text{ s}; \; R_{DC}/v = \tfrac{150}{80} = \tfrac{15}{8} \text{ s}$$

Therefore

$$y_B = 0.72 \sin [40\pi (t + \tfrac{5}{8})]$$

and

$$y_C = 0.14 \sin [40\pi (t - \tfrac{15}{8})]$$

Example 2.3

Two particles B and D begin executing SHM at the same instant, with the same phase and in the same direction, with amplitudes of 8 cm and 6 cm and periods 2 s and 1 s, respectively. The plane waves sent out by each particle disturb a third particle C, which is 20 cm from D and 60 cm from B. If the speed of the waves is 40 cm s^{-1} find the resultant displacement of C 5 s after the particles start to vibrate.

The term 'plane' refers to the shape of the wavefront and indicates that the amplitude of the progressive waves emitted by the two particles stays constant throughout free space. Once the wave equations have been set up at C, the Principle of Superposition will be used to determine the resultant displacement.

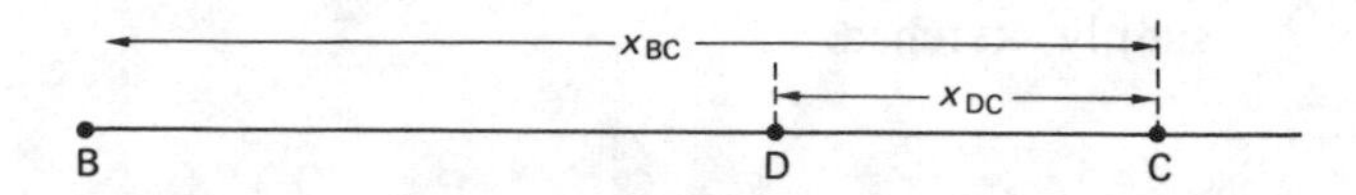

Data	*Given:*	$x_{DC} = 20$ cm $x_{BC} = 60$ cm $T_D = 1$ s $T_B = 2$ s $A_D = 6$ cm $A_B = 8$ cm $v = 40$ cm s^{-1}, $t = 5$ s
	Unknown:	initial phases at B and D, viz. ϕ_B and ϕ_D. y_C λ_D, λ_B
Relevant equations		$y = A \sin [(2\pi/\lambda)(x - vt) + \phi]$ $\lambda = v/f = vT$

We may write

$$(y_C)_D = A_D \sin [(2\pi/\lambda_D)(x_{DC} - vt) + \phi_D]$$
$$= A_D \sin [(2\pi/vT_D)(x_{DC} - vt) + \phi_D] \qquad (1)$$

and

$$(y_C)_B = A_B \sin [(2\pi/\lambda_B)(x_{BC} - vt) + \phi_B]$$
$$= A_B \sin [(2\pi/vT_B)(v_{BC} - vt) + \phi_B] \qquad (2)$$

By the Principle of Superposition

$v_C = (y_C)_D + (y_C)_B$

As ϕ_D and ϕ_B are not given, assume that they are both zero. Then

$$y_C = 6 \sin [(2\pi/40)(20 - 40 \times 5)] + 8 \sin [(2\pi/80)(60 - 40 \times 5)]$$
$$= 6 \sin (-9\pi) + 8 \sin (-7\pi/2)$$
$$= 0 + 8$$
$$= 8 \text{ cm}$$

Example 2.4

A point is subjected to two simple harmonic motions, given by

$y_1 = A \sin \omega t$
$y_2 = 2A \sin (\psi - \omega t)$

Determine the value of the ratio: maximum irradiance/minimum irradiance.

A phasor diagram will first be constructed from which a reduced vector diagram can be drawn. The resultant amplitude may then be calculated using the cosine law. As the irradiance is proportional to the square of the amplitude the ratio may be deduced.

Data	*Given:*	y_1 y_2
	Unknown:	ψ I
Relevant equations		$y_{RES} = y_1 + y_2$ $I_{RES} \propto A^2_{RES}$

Because of the way the progressive wave equation (2.2) has been defined it is first necessary to express y_1 as

$$y_1 = A \sin (\pi - \omega t)$$

Now a phasor diagram may be constructed to scale with radii in the ratio 1:2, or simply sketched.

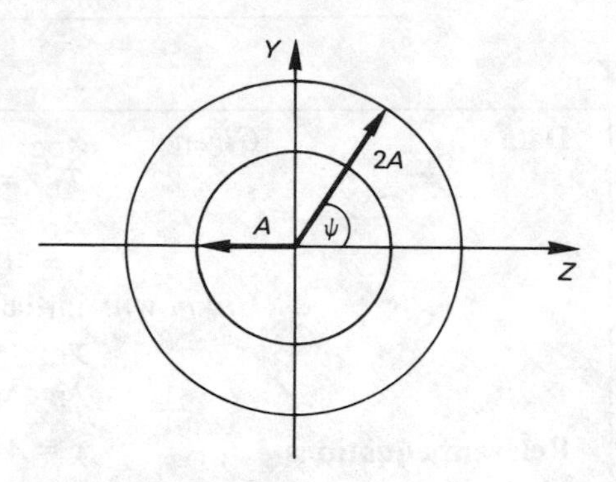

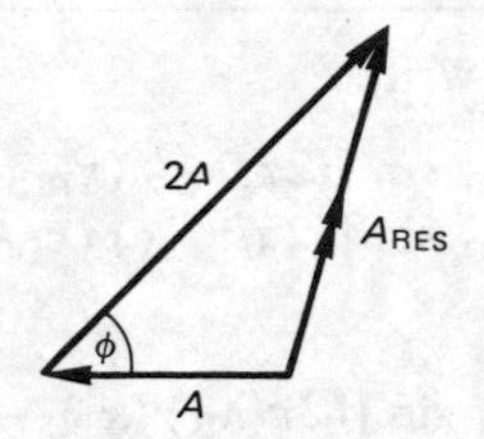

Applying the cosine rule, we obtain

$$A^2_{RES} = 4A^2 + A^2 - 4A^2 \cos \phi \quad (1)$$

Now, A^2_{RES} has its maximum value when $\cos \phi$ equals -1 and its minimum value when $\cos \phi$ equals $+1$, so

$$(A^2_{RES})_{max} = 9A^2 \quad (2)$$

and

$$(A^2_{RES})_{min} = A^2 \quad (3)$$

Therefore

$$I_{max} \propto 9A^2 \quad \text{or} \quad I_{max} = 9kA^2$$

and

$$I_{min} \propto A^2 \quad \text{or} \quad I_{min} = kA^2$$

where k is a constant of proportionality, which is the same in both cases. Hence

$$I_{max}/I_{min} = 9$$

NOTES

(i) Using the electromagnetic theory of radiation it can be shown that k is equal to $c\epsilon_o/2$, where c is the speed of light in free space and ϵ_o is the electric permittivity, and is equal to 8.85×10^{-12} F m^{-1}.

(ii) y_1 must be written in the form

$$y_1 = A \sin (\psi - \omega t)$$

in order that it has the same form as the general equation for progressive waves, viz.

$$y = A \sin [(2\pi/\lambda)x + \phi - \omega t)]$$

(see Section 2.4).

Example 2.5

A particle is acted on simultaneously by two simple harmonic motions:

$$y_1 = 4 \sin \left(\omega t - \frac{\pi}{4} \right) \qquad y_2 = 5 \sin \left(\omega t - \frac{\pi}{6} \right)$$

Find the amplitude of the resultant wave motion and write down the equation of the resultant SHM.

As in Example 2.4, each of the wave equations must be written in the form

$$y = A \sin (\psi - \omega t)$$

then they can be depicted on a phase-amplitude diagram. The radii of the circles will be 4 units and 5 units. The equation of the resultant wave motion may be represented as

$$y_{RES} = A_{RES} \sin (\omega t - \alpha)$$

Data	*Given:*	y_1
		y_2
	Unknown:	A_{RES}
		y_{RES}
		α
Relevant equations		$y = A \sin (\psi - \omega t)$
		$y_{RES} = y_1 + y_2$

Express y_1 in the form

$$y_1 = A_1 \sin (\psi_1 - \omega t)$$

i.e.

$$y_2 = 5 \sin \left(\frac{5\pi}{4} - \omega t \right)$$

In the same way

$$y_2 = 5 \sin \left(\frac{7\pi}{6} - \omega t \right)$$

Now construct the phasor diagram:

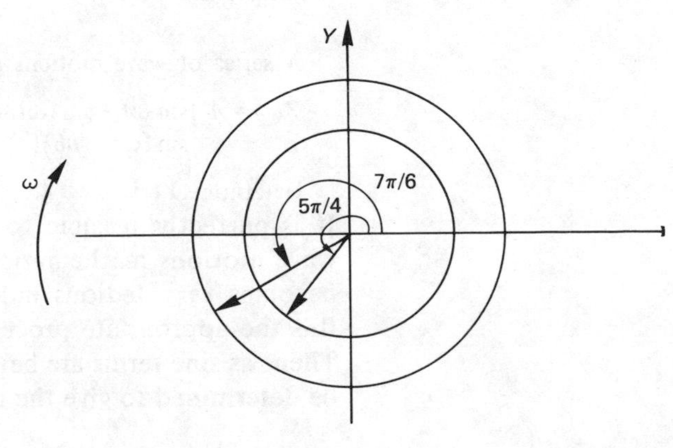

To find the resultant amplitude draw a vector diagram to scale (1 cm ≡ 1 unit is convenient).

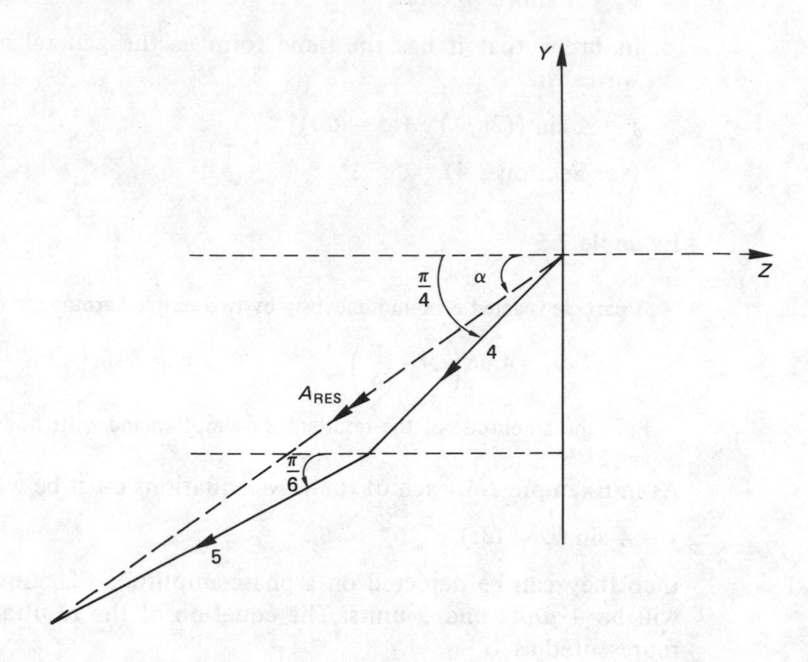

Direct measurement leads to A_{RES} = 8.9 units and $\alpha = 36.0° = \frac{\pi}{5}$ radian.

And so:

$$y_{RES} = 8.9 \sin\left(\omega t - \frac{\pi}{5}\right)$$

NOTES

(i) The phasor diagram depicted here relates the displacement of the wave to the phase angle at one instant of time only. The curved arrow outside the diagram indicates that both phasors rotate with angular velocity ω as time proceeds, but they keep the same angle between them.

(ii) If the waves were not plane-polarised, then a technique similar to that given in Example 5.1 would need to be adopted.

(iii) Electrical waves having the same basic sinusoidal form can be treated in this way.

Example 2.6

A series of wave motions are added together to give a resultant displacement $Y(t)$, where

$$Y(t) = A\ [\sin \omega t + \sin(\omega t - \delta) + \sin(\omega t - 2\delta) + \ldots + \sin(\omega t - n\delta)]$$

Determine $Y(t)$.

It is perfectly feasible to find the resultant displacement of two, or even three, wave motions mathematically by expanding the sine terms but after that the task becomes very tedious indeed and mistakes can be easily made. In examples like this the appropriate procedure is to express the phase term as a complex number. Then, as sine terms are being added, the imaginary part of the final expression can be determined to give the desired solution.

Data	*Given:*	The amplitude A is the same for each wave motion There is a constant phase difference δ between adjacent pairs of waves.
	Unknown:	$Y(t)$
Relevant equation		$y = A \sin(\omega t - r\delta) = A \text{ Im} \exp \mathrm{i} [\omega t - r\delta]$ (where Im stands for the imaginary part of the complex number).

$$\begin{aligned} Y(t) &= A \text{ Im} \exp(\mathrm{i}\omega t) [1 + \exp(-\mathrm{i}\delta) + \exp(-\mathrm{i}2\delta) + \ldots + \exp(-\mathrm{i}n\delta)] \\ &= A \text{ Im} \exp(\mathrm{i}\omega t) [1/[1 - \exp(-\mathrm{i}\delta)]] \\ &= MA \text{ Im} \exp[\mathrm{i}(\omega t + \phi)] \qquad (1) \end{aligned}$$

where

$1/[1 - \exp(-\mathrm{i}\delta)] = M \exp(\mathrm{i}\phi)$ (M is the modulus of the complex number)

The irradiance is given by

$$\begin{aligned} I &\propto (\text{amplitude})^2 \\ &\propto M^2 A^2 \end{aligned}$$

Now

$$M^2 = M \exp(\mathrm{i}\phi) \times M \exp(-\mathrm{i}\phi) \qquad (2)$$

N.B. The product of a complex number with its complex conjugate is a real number. Therefore

$$\begin{aligned} M^2 &= 1/[1 - \exp(-\mathrm{i}\delta)] \times 1/[1 - \exp(\mathrm{i}\delta)] \\ &= 1/[1 - (\exp(-\mathrm{i}\delta) + \exp(\mathrm{i}\delta)) + 1] \\ &= 1/2(1 - \cos\delta) \\ &= 1/[4 \sin^2(\delta/2)] \end{aligned}$$

Hence

$$M = 1/[2 \sin(\delta/2)] \qquad (3)$$

and the resultant displacement $Y(t)$ can be written

$$\begin{aligned} Y(t) &= A/[2 \sin(\delta/2)] \text{ Im} \exp[\mathrm{i}(\omega t + \phi)] \\ &= A/[2 \sin(\delta/2)] \sin(\omega t + \phi) \qquad (4) \end{aligned}$$

NOTES

(i) The expression for the resultant displacement $Y(t)$ has the same form as the displacement of the individual waves.

(ii) This kind of analysis is used to derive the Airy irradiance distribution of the transmitted wave system in multiple-beam interferometry.

Example 2.7

To repeat Example 2.5 using complex numbers.

As explained in Example 2.6, each simple harmonic motion can be written in complex form as

$$y_1(t) = 4 \exp\left[\mathrm{i}\left(\omega t - \frac{\pi}{4}\right)\right]$$

$$y_1(t) = 5 \exp\left[\mathrm{i}\left(\omega t - \frac{\pi}{6}\right)\right]$$

Here $\mathscr{y}(t)$ is the complex displacement. The resultant motion may then be expressed as $M \exp(i\phi)$, from which the resultant displacement may be found by taking the imaginary part of the exponential.

Data	*Given:*	$y_1(t) = 4 \sin\left(\omega t - \frac{\pi}{4}\right)$
		$y_2(t) = 5 \sin\left(\omega t - \frac{\pi}{6}\right)$
	Unknown:	y_{RES}
Relevant equation		$y(t) = A \sin(\omega t - \theta) = A \text{ Im} \exp[i(\omega t - \theta)]$

$$y_1 + y_2 = 4 \text{ Im} \exp\left[i\left(\omega t - \frac{\pi}{4}\right)\right] + 5 \text{ Im} \exp\left[i\left(\omega t - \frac{\pi}{6}\right)\right]$$

$$= \text{Im} \exp(i\omega t)\, [4 \exp(-i\pi/4) + 5 \exp(-i\pi/6)]$$

Write the term in [ ] as $M \exp(i\phi)$ so that

$$y_{\mathrm{RES}} = \text{Im } M \exp[i(\omega t + \phi)]$$

Now M multiplied by its complex conjugate M^* is a real number so

$$\begin{aligned} MM^* = M^2 &= [4 \exp(-i\pi/4) + 5 \exp(-i\pi/6)]\ [4 \exp(i\pi/4) + 5 \exp(i\pi/6)] \\ &= 41 + 20\, [\exp(-i\pi/12) + \exp(i\pi/12)] \\ &= 41 + 40 \cos(\pi/12) \end{aligned} \quad (3)$$

giving

$M = 8.9$ units

Therefore

$$y_{\mathrm{RES}} = 8.9 \text{ Im} \exp[i(\omega t + \phi)]$$

ϕ can be found from the real and imaginary parts of the expression contained within [ ] in (1). Using de Moivre's theorem this expression can be written

$$4\left(\cos\frac{\pi}{4} - i \sin\frac{\pi}{4}\right) + 5\left(\cos\frac{\pi}{6} - i \sin\frac{\pi}{6}\right)$$

Collecting real and imaginary terms together gives

$$(2\sqrt{2} + 5\sqrt{3}/2) - i\,(2\sqrt{2} + 5/2)$$

Thus

$$\tan\phi = -\,[2\sqrt{2} + 5/2]/[2\sqrt{2} + 5\sqrt{3}/2]$$

and

$$\begin{aligned} \phi &= -\tan^{-1}\,[5.328/7.158] \\ &= -36° \end{aligned}$$

Now we can write

$$\begin{aligned} y_{\mathrm{RES}} &= 8.9 \text{ Im} \exp[i(\omega t - 36°)] \\ &= 8.9 \sin(\omega t - 36°) \end{aligned}$$

which agrees with the earlier result.

Example 2.8

The refractive index of a glass prism satisfies the Cauchy relation

$$n = 1.5700 + 1.000 \times 10^4/\lambda^2$$

where λ is in nm.

Calculate the phase and group velocity at a wavelength of 600 nm. Take the velocity of light in free space to be $2.9979\,2 \times 10^8$ m s^{-1}.

In free space the velocity of light has a well-defined value but in a dispersive medium, such as glass, the velocity varies with wavelength (see NOTE (iv) to Example 1.1). Monochromatic light is pure fiction. If it existed its wave train would be infinitely long. All light signals observed in nature are described in terms of a wave train of finite length which is composed of an infinite number of component sinusoidal waves of various wavelengths; the shorter the wave train the larger the range of wavelengths involved, and vice versa.

In a dispersive medium each component wave moves with a slightly different velocity. They appear to be generated at one end of the train and disappear at the other. The velocity of these component waves is called the *phase* velocity v and the velocity of the train is the *group* velocity u.

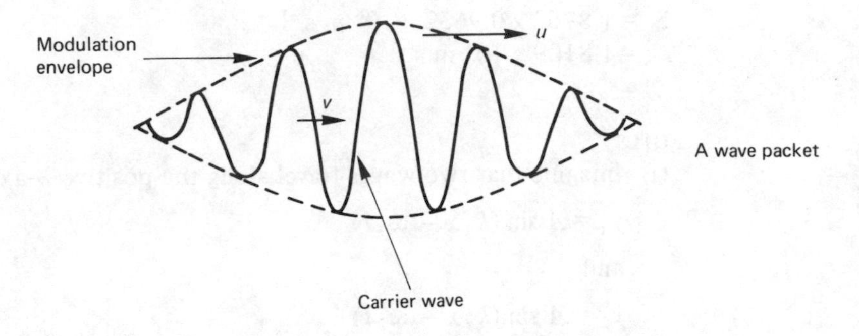

Data	*Given:*	Cauchy coefficients	$L = 1.5700$ $M = 1.000 \times 10^4$ nm^2 $\lambda = 600$ nm
	Unknown:	v u $\mathrm{d}v/\mathrm{d}\lambda$	
Relevant equation		$u = v - \lambda \mathrm{d}v/\mathrm{d}\lambda$	

The phase velocity is simply c/n.

We can express $\mathrm{d}v/\mathrm{d}\lambda$ as

$$\mathrm{d}v/\mathrm{d}\lambda = (\mathrm{d}v/\mathrm{d}n)\,(\mathrm{d}n/\mathrm{d}\lambda) \qquad (1)$$

So, as

$$\mathrm{d}v/\mathrm{d}n = -\,c/n^2$$

and

$$\mathrm{d}n/\mathrm{d}\lambda = -\,2B/\lambda^3$$

we have

$$\mathrm{d}v/\mathrm{d}\lambda = 2cM/n^2\lambda^3 \qquad (2)$$

Hence

$$u = (c/n)\,(1 - 2M/n\lambda^2) \qquad (3)$$

Now

$$M/\lambda^2 = 1.000 \times 10^4/3.600 \times 10^5$$
$$= 0.0278$$

and

$$n = 1.5700 + 0.0278$$
$$= 1.5978$$

Hence

$$v = c/n$$
$$= 2.99792 \times 10^8/1.5978$$
$$= 1.8762 \times 10^8 \text{ m s}^{-1}$$

and

$$M/n\lambda^2 = 0.0278/1.5978$$
$$= 0.0174$$

$$\therefore u = 1.8762 \times 10^8 (1 - 0.0348)$$
$$= 1.8762 \times 0.9652 \times 10^8 \text{ m s}^{-1}$$
$$= 1.8109 \times 10^8 \text{ m s}^{-1}$$

NOTES

(i) Imagine that two waves travel along the positive X-axis with equations

$$y_1 = A \sin (k_1 x - \omega_1 t)$$

and

$$y_2 = A \sin (k_2 x - \omega_2 t)$$

The resultant wave has a displacement y given by

$$y = 2A \sin \left\{ \frac{(k_1 + k_2)}{2} x - \frac{(\omega_1 + \omega_2)}{2} t \right\} \cos \left\{ \frac{(k_1 - k_2)}{2} x - \frac{(\omega_1 - \omega_2)}{2} t \right\}$$

$$= [2A \cos (k_\text{m} x - \omega_\text{m} t)] \sin (\overline{k}x - \overline{\omega}t)$$

where $\overline{k}$ and $\overline{\omega}$ are mean values, k_m is the modulation propagation number and ω_m is the modulation frequency.

The resultant wave is still a progressive wave because it has the same form as the individual waves. However, its amplitude is no longer constant. It has a cosine variation. This is an example of *amplitude modulation*. Detailed analysis shows that the phase velocity is given by

$$v = \overline{\omega}/\overline{k}$$

and the group velocity by

$$u = \omega_\text{m}/k_\text{m} = (\omega_1 - \omega_2)/(k_1 - k_2)$$

If the bracketed terms are small, then we can write

$$u = \text{d}\omega/\text{d}k$$

Since

$$\omega = vk$$

differentiation leads to

$$u = v + k \,\text{d}v/\text{d}k$$
$$= v - \lambda \,\text{d}v/\text{d}\lambda$$

(ii) All velocity of light experiments measure the group velocity.

(iii) Information propagated as a modulated wave cannot be transmitted at speeds greater than c.

(iv) It is possible to define a 'group' refractive index n_g as

$$n_g = c/u$$

(v) The phase velocity can be greater than c, as may occur when X-rays are beamed on to a glass plate.

Example 2.9

The resonant cavity of a He-Ne laser is 1.50 m long. The spectral line width of the emitted radiation is 1500 MHz. Determine (a) the number of axial modes that will be generated for a central wavelength of 638.2 nm, and (b) their wavelength difference. Assume that the cavity mirrors are plane and the speed of light in the gas mixture is 3.00×10^8 m s^{-1}.

The axial (or longitudinal) modes of the resonant cavity are stationary wave systems. Nodes are formed at the mirrors. So if the cavity length is d, then the condition that a stationary wave will be generated is

$$d = m\lambda/2$$

In laser work, m is generally called the *mode designation* but it is equivalent to the order of interference. As m can only be integral the wavelength difference between modes can be found by differentiating the above expression.

The atoms or molecules of a gas are constantly moving with a random distribution of velocities. If an atom (or molecule) moves along the line-of-sight of a detector during the period when it is de-exciting then the wavelength of the emitted radiation will be shifted by a certain amount. So, if an atom *at rest* with respect to the detector emits radiation of wavelength λ_o then one moving with velocity v_x emits radiation having an apparent wavelength λ', given by

$$\lambda' = \lambda_o\ (1 \pm v_x/c)$$

Here, the line-of-sight of the detector is designated the X-axis; the negative sign is taken if the atom is moving towards the detector, and the positive sign if away from it. This is an example of the familiar *Doppler* effect. Other atoms will possess a different velocity along this X-direction with the result that a range of wave-

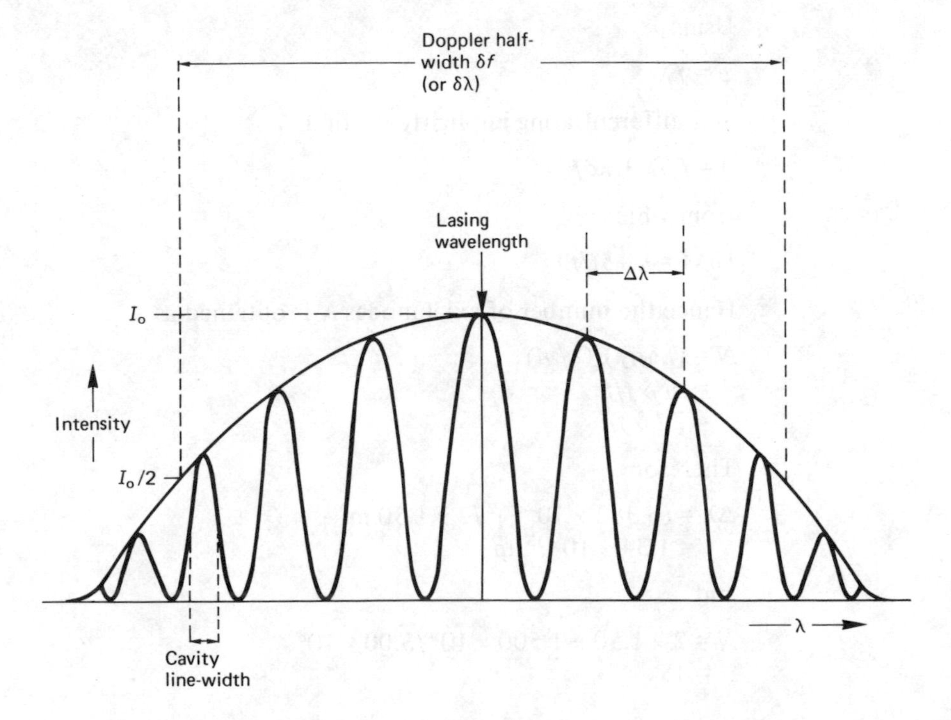

lengths will be detected, and the spectral line is Doppler broadened. The distribution of velocities in a perfect gas obeys a Gaussian equation of the form

$$f(v_x)\,\mathrm{d}v_x = C \exp(-mv_x^2/2kT)\,\mathrm{d}v_x$$

Data	*Given:*	Doppler half-width $\delta f = 1.500 \times 10^9$ Hz cavity length $d = 1.50$ m speed of light $c = 3.00 \times 10^8$ m s^{-1}
	Unknown:	mode separation $\Delta\lambda$ number of modes N
Relevant equations		$d = m\lambda/2$ $N = \delta\lambda/\Delta\lambda$

As

$$d = m\lambda/2 \tag{1}$$

differentiation leads to

$$0 = \Delta m\lambda/2 + m\Delta\lambda/2$$

which gives

$$\Delta\lambda = -\Delta m\lambda/m \tag{2}$$

We will omit the negative sign because we are only interested in the absolute value of $\Delta\lambda$. Further, as m can take integral values only, Δm must, therefore, be 1. Hence, the wavelength difference between modes is

$$\Delta\lambda = \lambda/m \tag{3}$$

or

$$\Delta\lambda = \lambda^2/2d \tag{4}$$

Using

$$c = f\lambda$$

and differentiating implicitly we find

$$0 = f\,\delta\lambda + \lambda\delta f$$

from which

$$|\delta\lambda| = \lambda\,|\delta f/f| \tag{5}$$

Hence the number of axial modes N is obtained as

$$\begin{aligned} N &= (\lambda\delta f/f)/(\lambda/m) \\ &= m\,\delta f/f \\ &= 2d\,\delta f/c \end{aligned} \tag{6}$$

Therefore

$$\begin{aligned} \Delta\lambda &= (6.382 \times 10^{-7})^2/2 \times 1.50 \text{ m} \\ &= 1.34 \times 10^{-13} \text{ m} \end{aligned}$$

and

$$\begin{aligned} N &= 2 \times 1.50 \times 1.500 \times 10^9/3.00 \times 10^8 \\ &= 15 \end{aligned}$$

NOTES

(i) The Doppler half-width is the range of frequencies lying between points where the spectral intensity equals half the peak intensity.

(ii) To reduce the number of axial modes in the laser output either the cavity length should be decreased or the output must be filtered. The disadvantage of the former solution is that the intensity of the output falls.

(iii) In practice, transverse electromagnetic modes will be generated also.

2.9 Questions

2.1

(a) Prove that $y(x, t) = g(x - vt)$ is a solution of the general wave equation:

$$d^2y/dx^2 = (1/v^2)\, d^2y/dt^2$$

(b) Satisfy yourself that

$$y = a \sin (kx - \omega t)$$

is equivalent to $g(x - vt)$.

2.2

A plane wave is transmitted through free space with speed v along the x-direction.

(a) Determine how the phase of the wave at coordinate $x = x_1$ varies with time, and

(b) determine how the phase of the wave varies with x at a given instant of time t_1.

2.3

A transverse sinusoidal wave motion of wavelength 2.0 cm is propagated along a thin wire. The instantaneous displacement (in cm) at a given point on the wire is given by

$$y = 3.0 \sin (3t/\pi)$$

Determine the displacement at a distance of 2.0 cm to the left and right of this point.

2.4

A transverse sinusoidal wave is generated four times each second at one end of a horizontal string which moves vertically through a distance of 1.0 cm. If the speed of propagation is 15.0 cm s^{-1} find the equation of the wave at time t if it moves (a) from left to right and (b) from right to left.

You may assume that the coordinates of the end of the string are (0, 0) at time $t = 0$. Find the velocity and acceleration of a point on the string 5.0 cm from the end.

2.5

A plane wave has an amplitude of 5.0 cm and a periodic time of 1.0 s. It travels along the positive x-direction with a speed of 100 cm s^{-1}. At a certain instant of time the phase of a point M is $\pi/6$. Determine the phase of a point N lying a distance of 50.0 cm nearer to the source and of a point S 20.0 cm further away at the same instant.

2.6

Sum the two wave motions of Example 2.4 using trigonometric identities and complex number notation.

2.7

A source of spherical waves vibrates according to the relation

$$y = A(r) \sin (2\pi t/T)$$

where $A(r)$ is 2.0 cm at a distance of 20.0 cm from the source and the frequency is 2.0 Hz. If the waves travel at 160 cm s^{-1}, find the equation of motion of a particle in the wave train 760 cm from the source at time t.

2.8

Using a phasor diagram, determine the resultant wave motion of the following three wave motions at the same instant of time t:

$$3 \cos\left(\omega t + \frac{\pi}{6}\right) + 5 \cos\left(\omega t - \frac{\pi}{6}\right) + 2 \cos \omega t$$

2.9

Sum the wave motions given in Question 2.8 using (a) conventional algebraic analysis and (b) complex number analysis.

2.10

Electromagnetic waves include:

(a) acoustic waves;
(b) micro-waves;
(c) photons;
(d) radio-activity;
(e) none of these but . . .

2.11

An electric field disturbance is propagated from point to point *in vacuo*:

(a) instantaneously;
(b) at a speed of 0.3 Gm s^{-1};
(c) as a stationary wave;
(d) in the absence of a magnetic field;
(e) none of these but . . .

2.12

The general equation of a progressive wave:

(a) must have a constant amplitude term;
(b) has a dimensionless phase angle;
(c) contains a distance parameter only;
(d) enables the optical path difference to be found;
(e) none of these but . . .

2.13

The optical path difference between two waves:

(a) depends on the phase angle of each wave only;
(b) depends on the geometrical path difference only;
(c) can only be found if coherent light sources are used;
(d) is the product of the refractive index and the geometrical path difference;
(e) none of these but . . .

2.14

Monochromatic light:

(a) is made up of an infinite number of sinusoidal waves;
(b) has a phase velocity which is less than the group velocity;
(c) suffers a discontinuous change in phase between its component waves;
(d) must have existed in the past at time $-\infty$ and will continue to exist into the future at time $+\infty$;
(e) none of these but . . .

2.15

The resonant cavity of a He-Ne laser:

(a) emits radiation with a frequency of 1.5 MHz;
(b) will only allow lasing action to take place with plane mirrors;
(c) must have a short length to reduce the number of axial modes;
(d) must have a length which is a multiple of half wavelengths;
(e) none of these but . . .

2.16

Radio waves behave in a similar way to visible light waves because:

(a) they both propagate at the speed of light;
(b) are characterised by the same value of the electric permittivity;
(c) the wavelength does not matter;
(d) they are electromagnetic waves which obey Maxwell's equations;
(e) none of these but . . .

2.17

The Principle of Superposition:

(a) can only be applied to a maximum of two waves;
(b) requires a phase-amplitude diagram in order to calculate the resultant amplitude;
(c) requires a 'parallelogram of forces' concept to be applied in the calculation of the resultant amplitude;
(d) says that the resultant displacement of visible light waves only at a given point is the sum of the displacements of the individual waves;
(e) none of these but . . .

2.18

Spherical waves:

(a) obey the inverse square law;
(b) are emitted by a point source and therefore do not exist;
(c) have an amplitude which varies as $(\text{distance})^{-2}$;
(d) have an amplitude which varies as $(\text{distance})^{-1}$;
(e) none of these but . . .

2.19

Stationary waves:

(a) are really progressive waves which are not propagating through space;
(b) have a constant phase at a given spatial position for all times t;
(c) really have a wave equation of the form $f(x - vt)$ but they are propagating too quickly to be observed;
(d) are characterised by a series of nodes occurring at points $x = (2m + 1)\lambda/4$, where m is zero or an integer;
(e) none of these but . . .

2.10 Answers to Questions

2.1 (a) This is standard differentiation.
(b) $k = 2\pi/\lambda$; $\omega = 2\pi v/\lambda$

2.2 (a) $(\mathrm{d}y/\mathrm{d}t)_{x=x_1}$
(b) $(\mathrm{d}y/\mathrm{d}x)_{t=t_1}$

2.3 $y = 3.0 \sin [(3/\pi)(t + \frac{20}{3})]$
$y = 3.0 \sin [(3/\pi)(t - \frac{20}{3})]$

2.4 (a) $y = 0.05 \sin [8\pi (x/15 - t)]$
(b) $y = 0.05 \sin [8\pi (x/15 + t)]$
The velocity is $\mathrm{d}y/\mathrm{d}t = 0.4 \cos [8\pi (x/15 - t)]$
The acceleration is $\mathrm{d}^2 y/\mathrm{d}t^2 = 64\,\pi^2 y$

2.5 $-5\pi/6$ at N; $17\pi/30$ at S

Let the phase angle at M be

$(2\pi/\lambda)\ [(x - vt) + \phi] = \pi/6$

Then at N replace x by $(x - 50)$, whence the phase is $\left(-\pi + \frac{\pi}{6}\right)$ because the wavelength is 100 cm. Similarly for S.

2.6 (a) Express resultant displacement as

$$\begin{aligned} Y &= \sin \omega t\ [A - 2A \cos \psi] + \cos \omega t\ [2A \sin \psi] \\ &= B \sin \omega t + C \cos \omega t \\ &= (B^2 + C^2)^{\frac{1}{2}} \sin (\omega t + \phi) \end{aligned}$$

Show that

$(B^2 + C^2) = [A^2 + 4A^2 - 4A^2 \cos \psi]$, as given in Example 2.4

and

$\tan \phi = 2A \sin \psi / [A - 2A \cos \psi] = 2 \sin \psi / [1 - 2 \cos \psi]$

(b) Rewrite $A \sin \omega t$ as $A \sin (\pi - \omega t)$

Sum the expression in complex notation to give

$A\ \mathrm{Im} \exp (-\mathrm{i}\omega t)\ [\exp (\mathrm{i}\pi) + 2 \exp (\mathrm{i}\psi)]$

Equate with $MA \exp (\mathrm{i}\phi)$, and obtain

$$\begin{aligned} M^2 &= 1 + 4 + 2\ [\exp (\mathrm{i}\pi) \exp (-\mathrm{i}\psi) + \exp (-\mathrm{i}\pi) \exp (\mathrm{i}\psi)] \\ &= 5 - 2\ [\exp (-\mathrm{i}\phi) + \exp (\mathrm{i}\psi)] \\ &= 5 - 4 \cos \psi \end{aligned}$$

Follow the technique used in Examples 2.6 and 2.7 to obtain ϕ.

2.7 $y = 0.05 \sin [4\pi (t - \frac{19}{4})]$

Determine the time taken for the wave to travel 760 cm. The phase of the wave at this point at time t equals the phase at the source at the earlier time of $(t - 19/4)$. Also amplitude is inversely proportional to distance.

2.8 $y = 9.0 \cos \left(\omega t + \frac{\pi}{24}\right)$

Express each wave motion in the form: $y = A \sin (\psi - \omega t)$. Use the phasor technique of Example 2.5.

2.9 (a) Expand the cosine and sine terms. Collect together the $\cos \omega t$ and $\sin \omega t$ terms to give

$(4\sqrt{3} + 2) \cos \omega t + 1 \sin \omega t$

Now express as $M \cos (\omega t + \alpha)$; $\begin{cases} \alpha = \tan^{-1} [1/(4\sqrt{3} + 2)] \\ M = [(4\sqrt{3} + 2)^2 + 1]^{\frac{1}{2}} \end{cases}$

(b) Follow the technique used in Examples 2.6 and 2.7 and express the resultant wave motion as

$M \exp [\mathrm{i} (\omega t + \alpha)]$

to give the same result as (a).

2.10 (b)

2.11 (b)

2.12 (b)

2.13 (e)
If the waves are travelling in the same medium, and there is no phase change of π, then the

OPD = $n \times$ geometrical path difference

otherwise calculate individual optical path lengths plus any phase changes of π.

2.14 (e)
As there are *no* monochromatic light sources (d) cannot be correct.

2.15 (d)

2.16 (d)

2.17 (e)
Applies to all waves in the EM spectrum – expand definition in (d).

2.18 (d)

2.19 (d)

3 Interference

3.1 Reason for Studying Interference Phenomena

Basically, the reason for the study is to learn how interferometers operate and what quantities they are able to measure. There are two classes of interferometers: (i) those that study the properties of light itself, such as wavelength and spectral line width, and (ii) those that measure some quantity such as refractive index, displacement or surface topography. Some interferometers can perform both tasks. Historically, the Michelson interferometer is important because it established that the speed of light is independent of the Earth's motion – a tenet vital for the general framework of the Special Theory of Relativity. More recently, it has been involved in a study of the tidal effects of the Moon on the Earth and continental drift.

3.2 Coherence

The term *coherence* was first introduced in Section 2.4. There, it was used to define a particular attribute of two light sources, viz. their ability to generate a stationary fringe system. It is essential that the difference between their initial phases stays constant as time goes on.

The subject of coherence is highly mathematical at its highest level but it is possible to gain some understanding of its meaning at a lower qualitative level.

There are two classes of coherence: *temporal* and *spatial*. The latter, in fact, can be further sub-divided into *longitudinal* and *transverse* (or *lateral*) spatial coherence.

Take a small region of a light source – so small that it is effectively a point. Even so, however, it will consist of many atoms (and/or molecules). Each atom will become excited at some time or other and then de-excite with the emission of electromagnetic radiation. The atom will then stay in the quiescent state until it is re-excited once more. Each atom in the point will behave in a similar way except that it is impossible to predict exactly when the emission process will occur and what the precise wavelength of the emitted radiation will be. The waves emitted by the various atoms can be added together, or *Fourier-synthesised*, to give a resultant wave train generated by the 'point' source. It will look something like the one shown in Example 2.8. The 'point' source will continue to generate wave trains, but in passing from one wave train to the next there will be an abrupt change in phase – there is no way of predicting what the phase will be because of the random nature of the emission process.

A second small region of the light source will behave in a similar manner to the first one but will act independently from it. The two 'point' sources are said to be *mutually incoherent*. Here, of course, we are referring to the ordinary discharge lamps found in the laboratory. The laser, on the other hand, operates by having the radiation emitted by one 'point' region stimulating another to emit. In such cases the phases of the emitted radiation are highly correlated. This is not generally the case with ordinary discharge lamps. The resultant irradiance arising when wave trains from 'point' sources overlap is the sum of the individual irradiances. No interference fringes are produced. The only way interference will occur is if two or more wave trains are generated by a single 'point' source, e.g. using either a beam-splitter, such as the glass block in Fig. 3.2, or the Young's double slit arrangement.

Suppose that a wave train has been Fourier-synthesised from a range of wavelengths $\Delta\lambda$ generated by a 'point' source S. It can be characterised, therefore, by a mean wavelength $\bar{\lambda}$. Next, take two points P_1 and P_2 lying along the same direction but separated by a spatial distance L (see Fig. 3.1). There is a time delay of L/c between the wave train arriving at P_1 and at P_2. If $\Delta\lambda$ is very small the length of the wave train L_c is very long; in the limit as $\Delta\lambda \to 0$, $L_c \to \infty$ – we are dealing with monochromatic radiation which, due to its pure sinusoidal waveform, must have existed in the past at time $t = -\infty$ and will continue to exist into the future at $t = +\infty$. Monochromatic radiation is a figment of our imagination, although laser light comes close to attaining this idealised state.

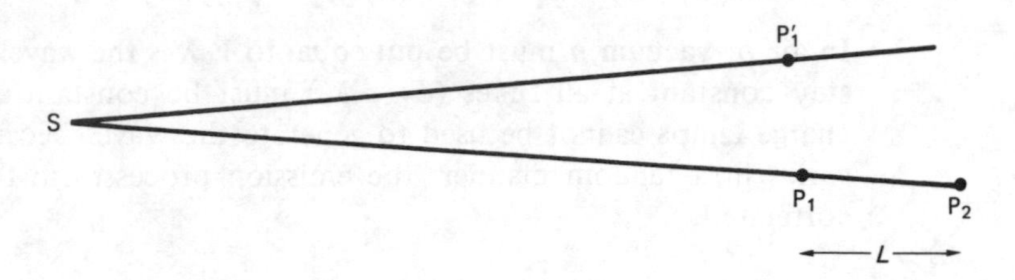

Figure 3.1

L_c is called the *coherence length* and L_c/c is called the *coherence time*. L_c is related to $\bar{\lambda}$ and the wavelength spread $\Delta\lambda$ by

$$L_c \simeq \lambda^2/\Delta\lambda \qquad (3.1)$$
$$\simeq c/\Delta f \quad \text{(in terms of frequency)}$$

The $\simeq$ sign means *about equal to*. It is not possible to give a precise value to L_c. For example, the light from a sodium discharge lamp has a coherence length of 2–3 cm whereas the cadmium red (589.4 nm) spectral line with a half-width of 0.65 pm has a coherence length of about 50 cm.

Returning to point P_1, we can say that when L_c is very long the phase there at different instants of time is highly correlated, i.e. the phase at time t'' can be related to the phase at an earlier time t'. There is said to be a high degree of *temporal* coherence. When L_c is short the degree of temporal coherence falls. The former will be characteristic of laser light and the latter of white light.

Similarly, as long as L_c is very much longer than L, the phase at P_2 can be related to the phase at P_1; there is a high degree of *longitudinal spatial* coherence. However, if L is greater than L_c there is little or no coherence.

If another point P_1' lies at the same radial distance from S as P_1 (Fig. 3.1) then a detector placed there will behave in exactly the same way as at P_1. There is a high degree of *transverse* (or *lateral*) spatial coherence between the optical fields at P_1 and P_1'.

Let us now take a Young's slits set-up and an extended source. If the slit secondary sources lie close to one another then they will receive light from the same part of the primary extended source. Fringes will be produced on a screen placed some distance away. What will occur if we have the facility for increasing the separation between the slits? We will then find that the slits begin to receive light from different 'points' in the extended source and the visibility of the fringe system diminishes until, eventually, the screen is uniformly illuminated. We have passed from a condition of complete coherence to one of complete incoherence. In the case of two mutually incoherent 'point' sources within the extended primary source which subtend an angle θ at the point midway between the slits the critical slit separation for this to occur is about equal to λ/θ. This critical separation is generally referred to as the transverse (or lateral) coherence width. With pin-holes, the coherence width is given by $1.22\lambda/\theta$. Thus in any Young's slits demonstration, or experiment, it is essential for the slit separation to be smaller than the coherence width.

3.3 Stationary Interference Patterns

The starting point of any discussion of interference is the relation for the phase difference δ between two waves travelling in the same direction but with different geometrical path lengths x_1 and x_2, viz.

$$\delta = (2\pi n/\lambda)\ [(x_2 - x_1) + (\lambda/2\pi n)\ (\phi_2 - \phi_1)] \tag{3.2}$$

In air or vacuum n must be put equal to 1. λ is the wavelength *in vacuo*. For δ to stay constant at all times $(\phi_2 - \phi_1)$ must be constant always. Two sodium discharge lamps cannot be used to generate the waves because $(\phi_2 - \phi_1)$ constantly varies in a random manner; the emission processes in the two lamps cannot be correlated.

(a) Division of Amplitude and Wavefront

There are two basic ways of satisfying (3.2). These are called *division of amplitude* and *division of wavefront*. The former is illustrated in Fig. 3.2.

To an observer it appears that the waves come from two sources S_1 and S_2 behind the glass block. As the waves are produced from the same incident wave the initial phases are the same. Hence δ will be constant. Thus S_1 and S_2 behave like coherent sources and the two reflected waves can interfere. (This is only strictly true with monochromatic light. With 'real' light sources t must not be too large. If this were the case the two wave trains would not overlap and interference would never occur.)

The interference conditions are:

$$\begin{aligned} 2n_g t \cos r &= (m + \tfrac{1}{2})\lambda \quad \text{– constructive interference} \\ 2n_g t \cos r &= m\lambda \quad \text{– destructive interference} \end{aligned} \tag{3.3}$$

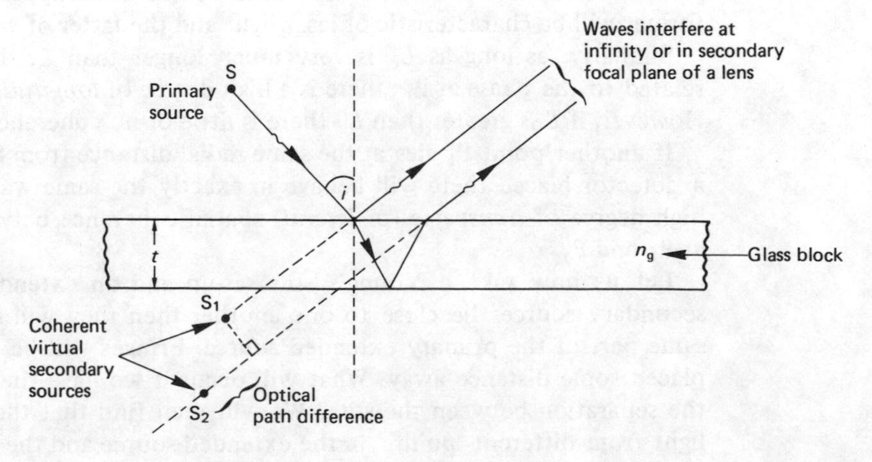

Figure 3.2

where m is the *order of interference*. The $\lambda/2$ term appears in the first condition because there is a phase change of π on reflection at the top surface of the block. There is a general rule to remember here: When light travels from a less optically dense medium towards a more optically dense *dielectric* medium and is reflected at the interface there is always a phase change of π. This is not the case with reflection at a metallic surface.

A family of interference fringes, characterised by different values of m, can be produced by allowing light to fall on the block at a variety of angles of incidence. Then certain of these angles will give rise to angles of refraction r which satisfy one or other of the interference conditions in (3.3). An extended light source is used for this purpose. The fringes are circular in three dimensions with the central fringe having the largest order of interference. The fringes are termed *fringes of equal inclination*.

Two-beam interference will also occur if the surfaces are inclined at a small angle to one another (~ 1 m radian), i.e. they form a wedge. The fringes will be linear and parallel to the line of intersection of the surfaces if the latter are perfectly flat. Experimentally, light from an extended source is usually incident normally on the wedge. A family of fringes, *localised* over a small region close to the wedge, is obtained because the wedge thickness is now the variable parameter. These are *fringes of equal thickness*.

Multiple interference will occur if the surfaces of the block are partially metallised to increase their reflectivity (see Figure 3.3). The transmitted system is normally used in experiments because the visibility of the fringe system is higher than with the reflected system. The Fabry–Perot etalon utilises this principle.

A very familiar example of Division of Wavefront is Young's double slit experiment. In this, a different portion of the expanding wavefront from a primary source passes through two rectangular slits or circular apertures in a screen. As these wavefronts spread out they overlap and interfere. A family of parallel bright and dark fringes are obtained with slits and a concentric ring pattern with circular apertures. The elementary theory of Young's slits assumes that the slits or apertures have no width so that diffraction effects can be neglected.

3.4 Fringe Visibility V

This is defined by the relation

$$V = (I_{max} - I_{min})/(I_{max} + I_{min}) \tag{3.4}$$

where I_{max} and I_{min}, respectively, are the irradiances of the bright and dark fringes. It can be shown that the visibility of the fringe system produced by two beams reflected from a parallel-sided glass block is very close to unity if a reflectivity of about 4% is assumed. On transmission, V is about 0.1. This means that it is very easy to distinguish the bright and dark fringes from one another in the reflection case but not in the transmission case, for then I_{max} and I_{min} are roughly equal. Fringe visibility is sometimes called *fringe contrast* or *modulation*.

3.5 Worked Examples

Example 3.1

The diameter of the tenth bright ring in a Newton's rings experiment decreases from 1.539 cm to 1.270 cm when a liquid is introduced between the lens and the parallel-sided block. Find the refractive index of the liquid if sodium yellow light is used of wavelength 589.3 nm.

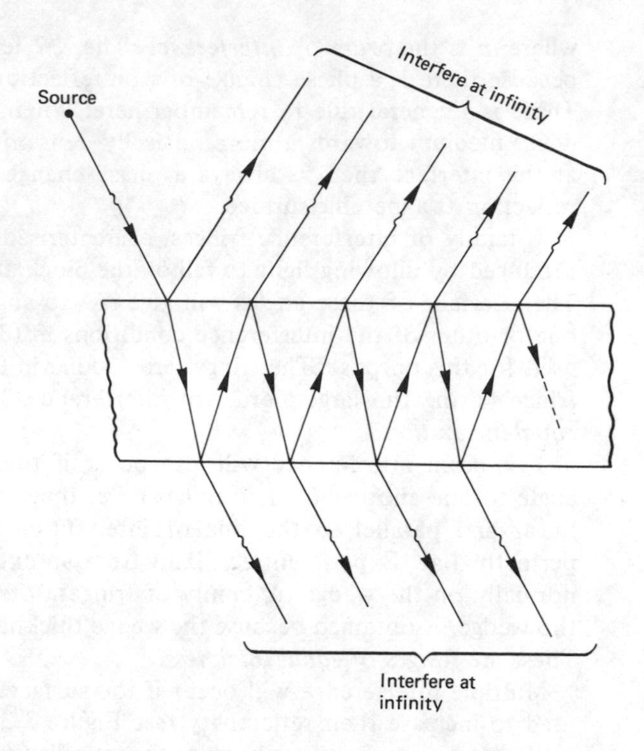

Figure 3.3

Two-beam interference occurs when light reflected from the bottom surface of the lens and the top surface of the block interferes. This region forms a wedge. Hence the fringes are called wedge fringes or fringes of equal thickness. In this case they form a set of concentric circular rings.

If the pole of the lens is in intimate contact with the block then the central fringe will be dark of order of interference $m = 0$. If this is not the case then the centre may be bright or dark of unknown order of interference. Each fringe is effectively a height contour, similar to those on a geographical map, and occurs at positions where the optical path difference between the two reflected wave systems is either $m\lambda$ or $(m + \frac{1}{2})\lambda$. It is essential to determine whether a phase change of π occurs on reflection at interfaces separating media of different optical densities; a phase change of π always occurs when light travels from a less optically dense medium towards a more optically dense dielectric medium and is partially or totally reflected at the interface. No phase change of π occurs if the more optically dense medium is metallic.

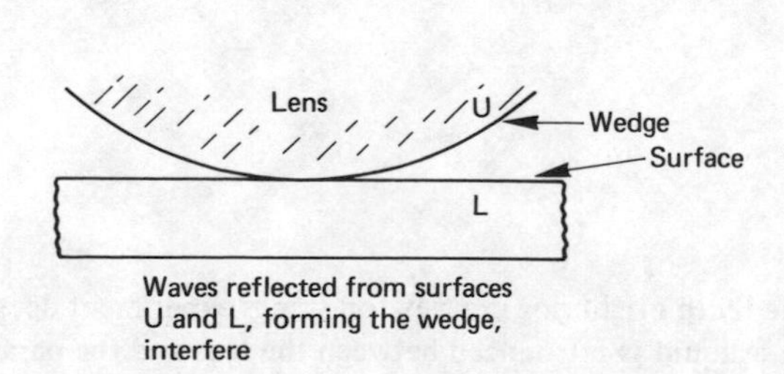

Waves reflected from surfaces U and L, forming the wedge, interfere

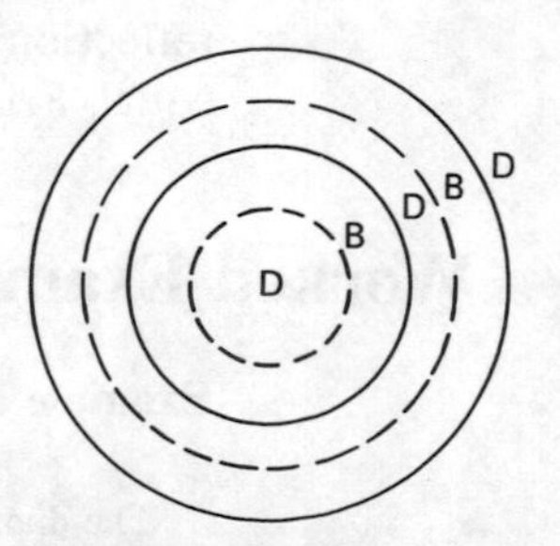

Concentric bright (B) and dark (D) ring system in an ideal system (with centre: dark $M = 0$)

Data	*Given:*	10th bright ring has $m = 9$
		$(d_{10})_{air} = 1.539$ cm
		$(d_{10})_{liq} = 1.270$ cm
		$\lambda = 5.893 \times 10^{-5}$ cm
	Unknown:	n_{liq}
		radius of curvature R of bottom lens surface
Relevant equations		$2nt = (m + \frac{1}{2})\lambda$
		$t(2R - t) = d^2/4$
		$d^2 = (2R\lambda/n)(2m + 1)$

We have

$$(d_{10}^2)_{air} = 2R\lambda\,(2m + 1) \qquad (1)$$

and

$$(d_{10}^2)_{liq} = (2R\lambda/n_{liq})\,(2m + 1) \qquad (2)$$

These give

$$\begin{aligned} n_{liq} &= (d_{10}^2)_{air}/(d_{10}^2)_{liq} \qquad (3) \\ &= (1.539/1.270)^2 = 2.368/1.613 \\ &= 1.468 \end{aligned}$$

NOTES

(i) The rings will be completely circular only if the lens is symmetric about its pole and its surface is spherical.

(ii) It is assumed that the lens is in intimate contact with the block so that the central fringe is dark. If there is a dust particle, or finger grease even, separating the lens from the block then the analysis is not straightforward. Suppose that the lens is a distance s above the block, then the condition for a bright fringe becomes

$$2n(t + s) = (N + m + \tfrac{1}{2})\lambda$$

where N is the unknown order of interference of the central fringe. Thus

$$d^2 = (4R\lambda/n)\,(N + m + \tfrac{1}{2}) - 8Rs$$

Now the ratio: $(d_{10}^2)_{air}/(d_{10}^2)_{liq}$ will not yield n_{liq}.

(iii) The question does not state that the light is incident normally on the wedge, but this is the usual experimental set-up.

(iv) If different light sources are used, n_{liq} can be found as a function of wavelength.

(v) The calculated value of n_{liq} suggests that the liquid is either glycerol or carbon tetrachloride. Whichever is correct will depend on the magnitude of the experimental errors – these are not stated in the question.

Example 3.2

A parallel-sided water film, 300 nm thick, is suspended horizontally in air. If it is illuminated with white light at normal incidence, what colour will it appear to be in reflected light? ($n_w = \frac{4}{3}$).

This is another example of two-beam interference; the reflected beams are from the top and bottom of the water film. Constructive interference occurs at one or more wavelengths in the visible range (350–700 nm) and satisfies the relation

optical path difference $= m\lambda$

As in Example 3.1, the optical path difference can only be determined once the number of phase changes of π are known. As both m and λ are unknown, a trial-and-error method must be adopted to determine λ. In this, various values of m are substituted into the above relation until the calculated wavelength lies in the visible region of the EM spectrum.

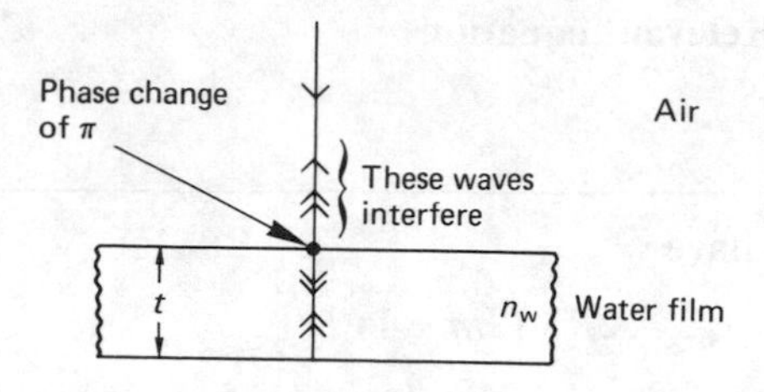

Data	*Given:*	$n_w = \frac{4}{3}$
		$t = 300$ nm
		angle of incidence = 0°
	Unknown:	order of interference m
		λ
Relevant equation		$2nt = (m + \frac{1}{2})\lambda$

In terms of wavelength, we have

$$\lambda = 2nt/(m + \tfrac{1}{2})$$

Try different values of m in this expression until λ lies between 350 and 700 nm. As

$$\lambda = (2 \times \tfrac{4}{3} \times 300)/(m + \tfrac{1}{2})$$
$$= 800/(m + \tfrac{1}{2})$$

construct a table for m and λ.

m	λ/nm
0	1600
1	533
2	320

λ equal to 533 nm is the only wavelength that will lead to constructive interference in the reflected light.

NOTES

(i) Some people do have the ability to detect wavelengths outside the visible range, but these are exceptional. The 'normal' eye has its highest sensitivity around 550 nm.

(ii) The visibility of the fringes observed in reflected light is higher than in transmitted light. For this reason experimental measurements are made with the former system.

Example 3.3

The reflectivity of crown glass ($n_g = 1.500$) is to be reduced by coating it with an appropriate thin film material. Determine (a) the material that should be used and (b) its metrical thickness.

The reflectivity R of a transparent material of refractive index n_g immersed in a medium of refractive index n_1 is defined as

$$R = (n_g - n_1)^2/(n_g + n_1)^2$$

for *normal* incidence.

In order to reduce R a material of refractive index n_f ($n_1 < n_f < n_g$) is deposited on the transparent substrate. If the optical thickness of the deposited film is $\lambda/4$, or $(2m + 1)\ \lambda/4$ in general, where $m = 0$ or an integer, then R should fall to zero. The coating is called an *anti-reflection* film. The metrical thickness is the actual, or as-deposited, thickness of the thin film and would be indicated by a quartz crystal oscillator or some other *in situ* thickness monitor.

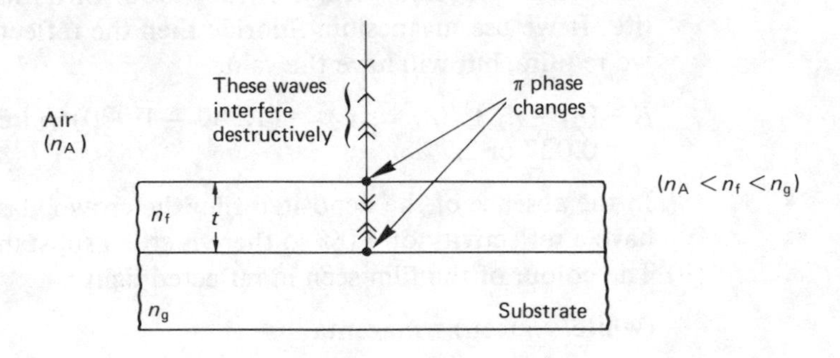

Data	*Given:*	$n_g = 1.500$
		$n_f t = \lambda/4$ or $(2m + 1)\ \lambda/4$
	Unknown:	n_f
		$R_{air/film}$
		$R_{film/glass}$
		λ
Relevant equations		$R_{air/film} = R_{film/glass}$ (ensures that the reflected beams have equal amplitudes)
		$2n_f t = (2m + 1)\lambda/2$ (anti-phase condition)

Now

$$(n_f - n_A)^2/(n_f + n_A)^2 = (n_g - n_f)^2/(n_g + n_f)^2 \tag{1}$$

$$\therefore n_f^2 = n_A n_g$$

or

$$n_f = (n_A n_g)^{\frac{1}{2}} \tag{2}$$

Thus

$$t = (2m + 1)\lambda/4n_f$$
$$= (2m + 1)\lambda/4(n_A n_g)^{\frac{1}{2}} \tag{3}$$

and

$$n_f = (1.000 \times 1.500)^{\frac{1}{2}}$$
$$= 1.225$$

For $m = 0$,

$t = \lambda/(4 \times 1.225)$

What value should λ have? The value usually chosen is the one to which the eye is most sensitive, i.e. about 550 nm, which lies in the greenish part of the visible spectrum.

Therefore

$t = 550/4.900$ nm
$= 112$ nm

Other values of t can be calculated for $m = 1, 2$, etc. These are all permissible.

NOTES

(i) No solid substance has a refractive index of 1.225. One material which has a refractive index close to this value, and which can be readily deposited, is magnesium fluoride with an n_f of 1.360. In its natural form it is called cryolite. If we use magnesium fluoride then the reflectivity will not fall to zero, as we require, but will have the value

$$R = (n_f - n_A)^2/(n_f + n_A)^2 = (1.360 - 1.000)^2/(1.360 + 1.000)^2$$
$$= 0.027 \text{ or } 2.7\%$$

In the absence of the deposited film the crown glass substrate can be shown to have a reflectivity of 4.0% so there is still a substantial reduction.

(ii) The colour of the film seen in reflected light is

(white – green) = magenta

This is the purplish colour which gives the 'bloom' to camera lenses.

Example 3.4

An optical system consists of two layers of magnesium fluoride ($n_{mf} = 1.360$) and three layers of titanium dioxide ($n_{td} = 2.450$) deposited alternately on a glass substrate ($n_g = 1.500$) with the titanium dioxide deposited first. Demonstrate that the system behaves like a high-reflection mirror if the optical thickness of each layer is $\lambda/4$ and light is incident normally in air.

Conceptually the problem is similar to Example 3.3. It is necessary to show that the beams reflected at the various interfaces are in phase with one another when they re-enter the air. Further, it is essential to determine whether any phase change of π occurs on reflection. If so, an extra optical path of $\lambda/2$ must be included in the overall optical path.

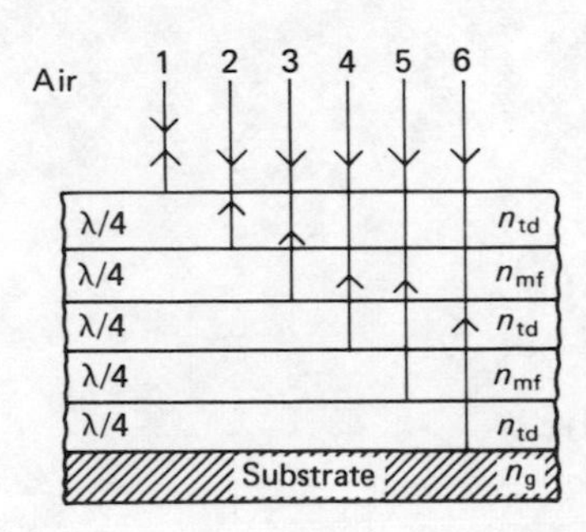

For clarity the incident beam has been divided into six individual beams

Data	Given:	$n_{mf} = 1.360$ (mf ≡ magnesium fluoride)
		$n_{td} = 2.450$ (td ≡ titanium dioxide)
		$n_g = 1.500$
		$i = 0°$
		optical thickness of each layer = $\lambda/4$
	Unknown:	phase difference between light reflected at the top surface and light reflected at each interface.

Table 3.1 gives the optical path difference between reflected wave 1 and a wave reflected at an interface within the stack. It can be seen that the optical path difference is $m\lambda$, where m is zero or an integer. Hence the reflected waves are all in phase and the stack acts as a high-reflection device.

Wave	*π phase change on reflection?*	*Optical PD*
1	yes	reference
2	no	0
3	yes	λ
4	no	λ
5	yes	2λ
6	no	2λ

NOTES

(i) The optical device is referred to as a multiple-layer dielectric stack, of the type that is used for the highly reflecting mirrors at the ends of the resonant cavity of a laser.

(ii) As the optical thickness of each layer is $\lambda/4$, this means that the device is designed to operate at one unique wavelength.

Example 3.5

A light source having two discrete wavelengths: $\lambda_1 = 600$ nm and $\lambda_2 = 450$ nm is used in a Newton's rings experiment. It is observed that the mth dark ring produced by λ_1 coincides with the $(m + 1)$th dark ring for λ_2. If the radius of curvature of the lens is 90.0 cm find the diameter of the mth dark ring for λ_1.

As with Example 3.1, two-beam interference occurs in an air-wedge formed by the bottom surface of the lens and the top surface of the supporting block. Each wavelength in the source produces its own ring system. Generally, the sodium yellow wavelengths are used in student experiments. These have values of 589.0 nm and 589.6 nm. If a high-resolution microscope is used to view the fringes two systems should be observed, but as the laboratory equipment is usually not of this standard one ring system only is seen. The diagram is similar to that in Example 3.1.

Data	Given:	$\lambda_1 = 600$ nm
		$\lambda_2 = 450$ nm
		order of interference for dark rings = $\begin{cases} m \text{ for } \lambda_1 \\ m+1 \text{ for } \lambda_2 \end{cases}$
		$R = 90.0$ cm
		wedge medium = air
	Unknown:	$(d_m)_{\lambda_1}$
		$(d_{m+1})_{\lambda_2}$
Relevant equations		$2nt = m\lambda$
		$d^2 = 4Rm\lambda/n$

We have

$$(d_m^2)_{\lambda_1} = 4R\lambda_1 m \quad (1)$$

and

$$(d_{m+1}^2)_{\lambda_2} = 4R\lambda_2(m+1) \quad (2)$$

In addition, because

$$(d_m)_{\lambda_1} = (d_{m+1})_{\lambda_2}$$
$$4R\lambda_1 m = 4R\lambda_2(m+1)$$

or

$$(m+1)/m = \lambda_1/\lambda_2 \quad (3)$$

Transposing (3) gives

$$1/m = (\lambda_1/\lambda_2) - 1$$

and

$$m = \lambda_2/(\lambda_1 - \lambda_2)$$

Therefore

$$(d_m^2)_{\lambda_1} = 4R\lambda_1\lambda_2/(\lambda_1 - \lambda_2) \quad (4)$$

and

$$(d_m^2)_{600} = 4 \times 90.0 \times 6.0 \times 10^{-5} \times 4.5 \times 10^{-5}/1.5 \times 10^{-5} \text{ cm}^2$$
$$= 6.5 \times 10^{-2} \text{ cm}^2$$

giving

$$(d_m)_{600} = 0.26 \text{ cm}$$

Example 3.6

Two thin glass plates are in contact at one edge and are separated at the other by a spacer 5 μm thick. The refractive indices of the upper and lower plates are 1.500 and 2.000, respectively. Fringes are observed in reflected light of wavelength 500 nm. (a) How many fringes are observed with air between the plates? (b) If the whole apparatus is immersed in oil of refractive index 1.800, how many fringes are now seen?

This is an example of two-beam wedge interference and the formation of fringes of equal thickness. The fringes will be linear and parallel to the line of intersection of the plates if they are both flat. We need to assume that this is the case here. If the plates are not flat then the fringes will deviate from linearity. The fringe at the line of intersection will be dark because a phase change of π occurs on reflection at the bottom plate.

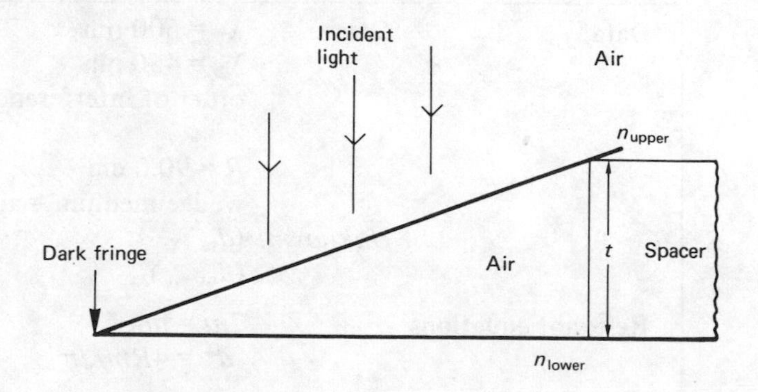

The plates are said to be thin so that interference in the plates themselves can be ignored. In any case, as light is incident on the wedge at one angle of incidence the same degree of interference will occur right across the field of view. This fringe of equal inclination will give rise to a uniform background (or 'noise') to the wedge fringes.

(a)

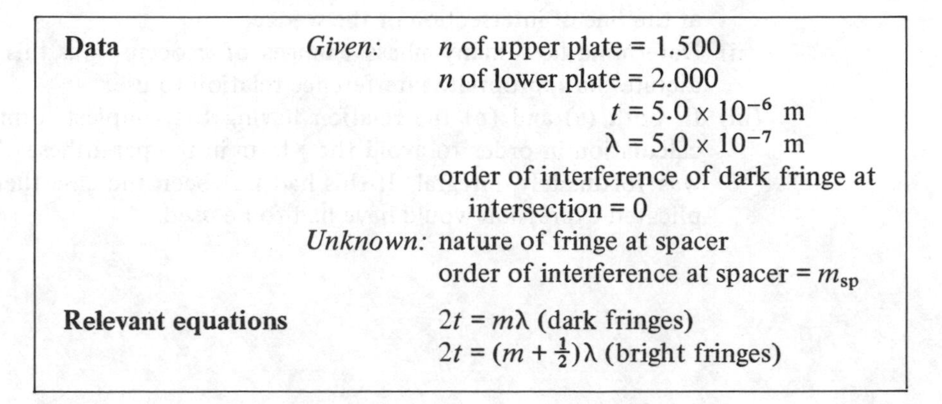

Data	*Given:*	n of upper plate = 1.500
		n of lower plate = 2.000
		$t = 5.0 \times 10^{-6}$ m
		$\lambda = 5.0 \times 10^{-7}$ m
		order of interference of dark fringe at intersection = 0
	Unknown:	nature of fringe at spacer
		order of interference at spacer = m_{sp}
Relevant equations		$2t = m\lambda$ (dark fringes)
		$2t = (m + \frac{1}{2})\lambda$ (bright fringes)

Suppose that there is a dark fringe at the spacer, then

$m_{sp} = 2t/\lambda$

This must be integral. If m_{sp} is non-integral then it means that the initial assumption is incorrect and that there is probably a bright fringe there.

The total number of dark fringes observed = 1 + m_{sp} and the total number of bright fringes = m_{sp} so

$$m_{sp} = 2 \times 5.0 \times 10^{-6}/5.0 \times 10^{-7} = 20$$

There are 21 dark fringes and 20 bright fringes.

(b)

The diagram is similar to that in Part (a) except that the wedge medium is oil, and there is now a bright fringe at the line of intersection.

Data	*Given:*	$n_{oil} = 1.800$
		$n_{upper} = 1.500$
		$n_{lower} = 2.000$
		$t = 5.0 \times 10^{-6}$ m
		$\lambda = 5.0 \times 10^{-7}$ m
	Unknown:	nature of fringe at spacer
		m_{sp}
Relevant equations		$2n_{oil}t = (m + \frac{1}{2})\lambda$ (dark; because there is now a phase change of π at the top surface of the wedge as well at the lower)
		$2n_{oil}t = m\lambda$ (bright)

Suppose, now, that there is a bright fringe at the spacer, then

$m_{sp} = 2n_{oil}\, t/\lambda$

As in (a), if this is integral then there are (m_{sp} + 1) bright fringes and m_{sp} dark.

m_{sp} = 2 x 1.800 x 5.0 x 10^{-6}/5.0 x 10^{-7}
= 36

So there are 37 bright fringes and 36 dark.

NOTES

(i) In such problems it is vital to determine the nature of the zero order fringe at the line of intersection of the wedge.

(ii) Determine how many phase changes of π occur – as this information will dictate the appropriate interference relation to use.

(iii) In both (a) and (b) the relation having the simplest form was used in the calculation in order to avoid the $\frac{1}{2}$ term in the parentheses. The value of m_{sp} was fortunately integral. If this had not been the case then the more complicated expression would have had to be used.

Example 3.7

A glass block (n_g = 1.500) is illuminated with light of wavelength 6.0 x 10^{-7} m. The reflected light enters a telescope focused at infinity. The axis of the telescope is perpendicular to the surfaces of the block. The objective lens has a focal length of 30.0 cm and a diaphragm, 1.0 cm in diameter, placed in its secondary focal plane limits the field of view to a central bright spot and to surrounding bright rings. Find the thickness of the block.

This is an example of two-beam interference and the production of fringes of equal inclination. To observe them the telescope must be focused at infinity. A parallel beam of light entering the telescope will interfere in the secondary focal plane of the objective.

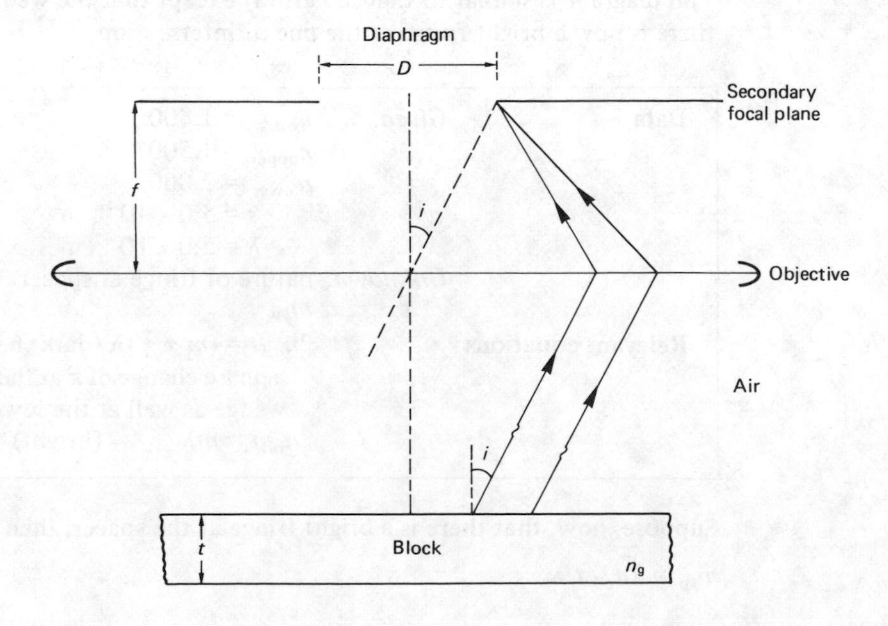

Data	Given:	$n_g = 1.500$
		$\lambda = 6.0 \times 10^{-5}$ cm
		$f = 30.0$ cm
		$D = 1.0$ cm
		No. of fringes = 10
	Unknown:	t
		order of interference m at centre of field of view
		angle of incidence i of light on block
		angle of refraction r
Relevant equations		$2n_g t \cos r = (m + \frac{1}{2})\lambda$ (bright)
		$2n_g t \cos r = m\lambda$ (dark)
		$n_g = \sin i / \sin r$

At the centre of the fringe system

$i = r = 0$; $\cos r = 1$

Suppose that the order of interference of the central bright spot is m, then

$$2n_g t = (m + \tfrac{1}{2})\lambda \quad (1)$$

At the periphery of the diaphragm,

$i \neq 0$, $\cos r \neq 1$

As the order of interference of the 10th bright ring is $(m - 10)$ this means that

$$2n_g t \cos r = (m - 10 + \tfrac{1}{2})\lambda \quad (2)$$

Subtract (2) from (1) to give

$$2n_g t\,(1 - \cos r) = 10\lambda \quad (3)$$

Then

$$t = 10\lambda/2n_g\,[1 - \cos r] \quad (4)$$

Further, as

$$\sin r = (1/n_g) \sin i$$

and

$$\cos r = [1 - (1/n_g)^2 \sin^2 i]^{\frac{1}{2}}$$

we find that

$$t = 10\lambda/2n_g\ [1 - [1 - (1/n_g)^2 \sin^2 i]^{\frac{1}{2}}] \quad (5)$$

As

$$\tan i = D/2f$$
$$= 0.5/30.0$$
$$= 1/60$$
$$\sin i = 1.6667 \times 10^{-2}$$

and

$$t = 10 \times 6.0 \times 10^{-5}/\{2 \times 1.500 \times [1 - (1 - 1.2346 \times 10^{-4})^{\frac{1}{2}}]\}$$
$$= 3.2 \text{ cm}$$

NOTES

(i) The same kind of procedure can be adopted if the central spot is dark.

(ii) The parallel wave system shown in the diagram comes to a focus at the periphery of the diaphragm. This focal position is found using a familiar technique with *thin* lenses in ray-tracing exercises. It is called the *oblique ray* construction. First draw a line through the optical centre of the lens parallel to the wave direction. The intersection of the line with the secondary focal plane will give the point where the waves interfere.

(iii) Only the waves reflected by the block are shown in the diagram in order to present essential information only.

Example 3.8

A mercury discharge lamp is placed immediately behind a rectangular slit and the light is filtered to allow the 546 nm wavelength to fall on a Fresnel biprism, 25.0 cm away. Interference fringes are produced on a screen 75.0 cm from the biprism. If the biprism has angles of 2.00° and a refractive index of 1.500, determine the separation between neighbouring fringes.

As the name suggests the biprism consists of two small-angle prisms, base-to-base. It is used as an alternative to Young's slits in Division of Wavefront interference experiments or demonstrations. Two coherent virtual secondary sources S_1 and S_2 are produced in the plane of the primary source P, as depicted in the diagram below. Interference occurs in the secondary focal plane of the objective of a micrometer eyepiece, which is used to scan across the fringes, or on a suitably-positioned screen (as in the diagram). The fringes can only be produced by light refracted by each half of the biprism.

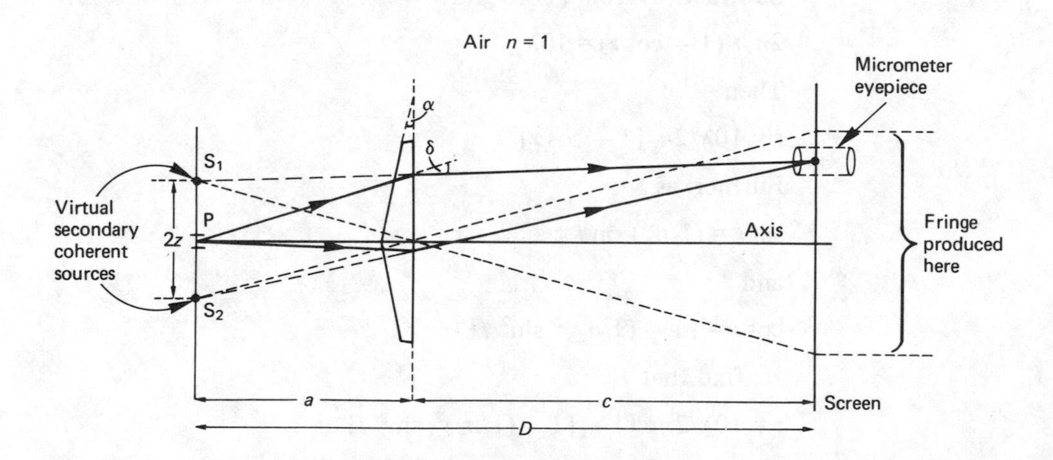

Data	*Given:*	refracting angle $\alpha = 2.00°$
		wavelength $\lambda = 546$ nm
		source–biprism distance $a = 25.0$ cm
		biprism–screen distance $c = 75.0$ cm
		refractive index n of biprism = 1.5000
	Unknown:	fringe separation s
		distance between virtual sources $2z$
Relevant equations		$\delta = (n - 1)\alpha$ for a thin prism
		$s = \lambda D/2z$ in air (the usual Young's slits expression)

The virtual source separation $2z$ can be expressed in terms of a and δ by using the small-angle approximation. Then

$$z = a\,\delta$$

and $2z$, in the conventional Young's slits expression given above, can be replaced by $2a(n-1)\alpha$. Thus

$$s = \lambda D/2a(n-1)\alpha$$

From the given data,

$$D = a + c = 1.00 \text{ m}$$

and

$$\alpha = 2.00 \times \pi/180 \text{ radians (for small angles)}$$

Hence

$$s = 5.46 \times 10^{-7} \times 1.00/[2 \times 0.500 \times 2.00 \times (\pi/180) \times 0.25] \text{ m}$$
$$= 1.97 \times 10^{-4} \text{ m}$$

NOTES

(i) For accurate measurements of the fringe separation it is more sensible to use the micrometer eyepiece.

(ii) For increased accuracy, it is essential to measure the separation between about 10 fringes or so and then obtain a mean separation.

(iii) The distance between the virtual secondary sources can be found by inserting a converging lens between the biprism and the micrometer eyepiece. However, it is essential that the source–eyepiece distance is made greater than four times the focal length of the lens. On moving the lens, two positions will be found in which a sharp image of the virtual sources can be obtained; in one position the distance between them is magnified d_1, say, and in the other it is reduced d_2, say. Then the true separation between them is the geometric mean of d_1 and d_2, i.e. $(d_1 d_2)^{\frac{1}{2}}$.

Example 3.9

A ship's mast is 25.0 m high. It is fitted with a radio transmitter operating at a wavelength greater than 3.00 m. A receiving station on land, 150.0 m above the sea, loses contact with the ship when the latter is 2.00 km away from it. Calculate a precise value for the operating wavelength.

The receiving station loses contact with the ship because the direct beam from the ship to the receiver interferes destructively with a beam reflected off the surface of the sea. The set-up is similar to a Young's slits arrangement. However care must be taken not to omit the phase change of π that occurs on reflection at the surface of the sea.

Destructive interference occurs when the optical path difference between the two beams is equal to $(m + \frac{1}{2})\lambda$, where once again m is the order of interference.

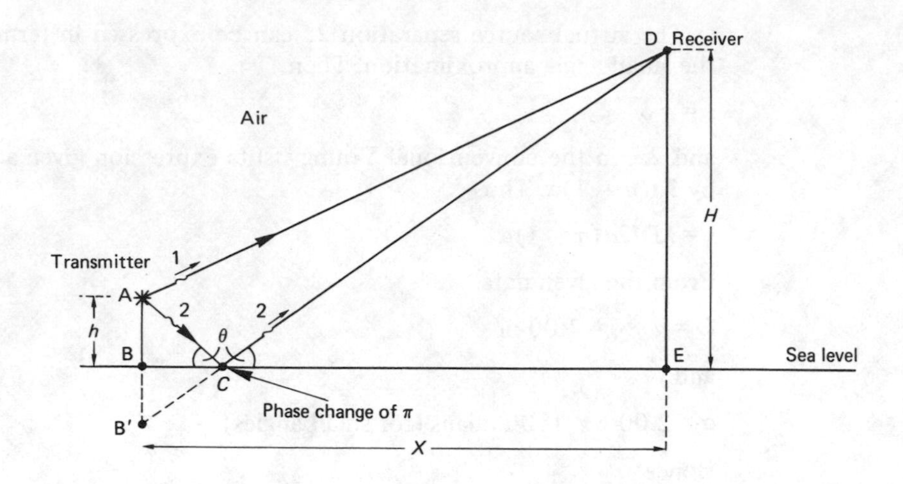

Data	*Given:*	h = 25.0 m X = 2.00 km H = 150.0 m
	Unknown:	λ (but > 3.00 m)
Relevant equations		optical path difference = $(m + \frac{1}{2})\lambda$ when receiver loses contact (optical path)$_1$ = [AD] (optical path)$_2$ = [AC + CD] + λ/2 = [B′D] + λ/2

There are two ways of calculating the optical path of wave 2:

(a) determine optical paths AC and CD separately;
(b) optical path AC is equal to optical path B′C. So to an observer at D the waves appear to come from two sources at A and B′.

Using (a):
As triangles ABC and DCE are similar, if we let BC equal x then CE equals $(X - x)$. Hence

$$AB/x = DE/(X - x)$$

and

$$x = AB \times X/(AB + DE) \tag{1}$$

then

$$[AD] = [(H - h)^2 + X^2]^{\frac{1}{2}} \quad \text{(because } n_A = 1\text{)} \tag{2}$$

and

$$[AC] = (h^2 + x^2)^{\frac{1}{2}}; \ [CD] = [H^2 + (X - x)^2]^{\frac{1}{2}}$$

i.e.

$$[ACD] = (h^2 + x^2)^{\frac{1}{2}} + [H^2 + (X - x)^2]^{\frac{1}{2}} + \lambda/2$$

The optical path difference may now be equated with $(m + \frac{1}{2})\lambda$, as in

$$(h^2 + x^2)^{\frac{1}{2}} + [H^2 + (X - x)^2]^{\frac{1}{2}} + \lambda/2 - [(H - h)^2 + X^2]^{\frac{1}{2}} = (m + \tfrac{1}{2})\lambda$$

which reduces to

$$(h^2 + x^2)^{\frac{1}{2}} + [H^2 + (X - x)^2]^{\frac{1}{2}} - [(H - h)^2 + X^2]^{\frac{1}{2}} = m\lambda \tag{3}$$

Using (b):

$$[\mathrm{B'D}] = [X^2 + (H+h)^2]^{\frac{1}{2}} \tag{4}$$

This is easier to handle than [ACD] although the geometrical path lengths are identical. So using it we have

optical path difference = $[\mathrm{B'D}] - [\mathrm{AD}] = (m + \frac{1}{2})\lambda$

giving

$$[X^2 + (H+h)^2]^{\frac{1}{2}} - [X^2 + (H-h)^2]^{\frac{1}{2}} = m\lambda \tag{5}$$

As X is very much larger than either H or h, use the Binomial expansion to give

$$X\,[1 + (H+h)^2/2X^2] - X\,[1 + (H-h)^2/2X^2] = m\lambda \tag{6}$$

or

$$2Hh/X = m\lambda$$

from which

$$\lambda = 2Hh/mX \tag{7}$$
$$= 2 \times 150.0 \times 25.0/m \times 2000$$

Try various values of m until $\lambda > 3.00$ m

For $m = 1$:

$\lambda = 3.75$ m

which satisfies the required condition.

NOTES

(i) This example demonstrates how a modification of the initial diagram leads to an enormous simplification in the mathematics.

(ii) The example also demonstrates the need to take extra care in not omitting to include the phase change of π that occurs at the surface of the sea. In the usual Young's slits experiment this extra term would not exist.

Example 3.10

Calculate the half-width of multiple-beam interference fringes if the coefficient of reflectance of the Fabry-Perot etalon plates is 0.90.

The etalon glass plates are partially metallised on their inner surfaces. The high reflectance leads to a large number of waves being reflected and transmitted, although, experimentally, the latter system of waves is used due to the higher visibility of the fringes. Airy's formula relates the resultant irradiance I_{T} of the transmitted wave system with the amplitude A of the incident light, the reflectance R (or transmittance T) of the plates and the phase difference δ between adjacent pairs of transmitted waves. Constructive interference occurs for $\delta = 0, \pm 2\pi, \pm 4\pi, \ldots, \pm 2m\pi$. At some value of δ, call it δ', the irradiance will fall to $(I_{\mathrm{T}})_{\mathrm{max}}/2$. The value $2\delta'$ is then referred to as the half-width of the fringes.

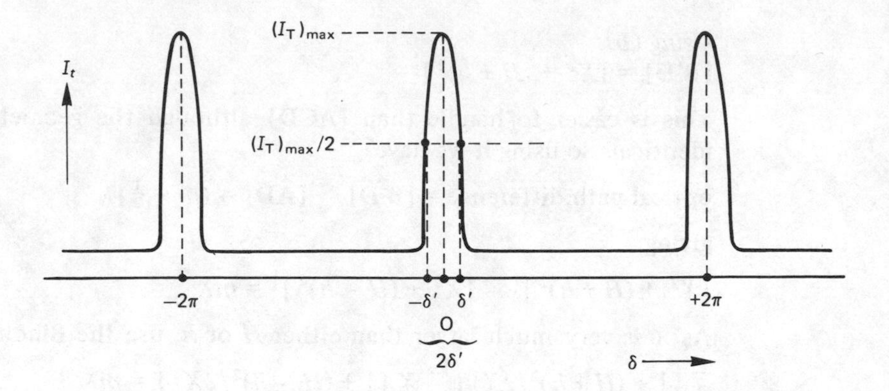

Data	*Given:*	$R = 0.90$
	Unknown:	$(I_T)_{max}$
		etalon spacing d
		n_{medium}
		angle of incidence
		δ'
Relevant equations		$I_T = (I_T)_{max}/[1 + 4R \sin^2 (\delta/2)/(1 - R)^2]$
		$\delta = (2\pi/\lambda).2n_{med}d \cos\theta$

At $\delta = 0$:

$$I_T = (I_T)_{max} \tag{1}$$

At $\delta = \delta'$:

$$I_T = (I_T)_{max}/2 \tag{2}$$

$$= (I_T)_{max}/[1 + 4R \sin^2 (\delta'/2)/(1 - R)^2] \tag{3}$$

Dividing (2) by (3) gives

$$1 = 4R \sin^2 (\delta'/2)/(1 - R)^2 = 2 \tag{4}$$

and

$$\delta' = 2 [\sin^{-1} ((1 - R)^2/4R)^{\frac{1}{2}}]$$

The half-width is

$$2\delta' = 4 [\sin^{-1} ((1 - R)^2/4R)^{\frac{1}{2}}] \tag{5}$$

$$= 4 [\sin^{-1} ((1 - 0.90)^2/3.60)^{\frac{1}{2}}]$$
$$= 4 [\sin^{-1} (1.00 \times 10^{-2}/3.60)^{\frac{1}{2}}]$$
$$= 12.1^\circ$$

Example 3.11

Multiple-beam fringes are projected on to a screen using a lens of focal length equal to 1.00 m. Calculate the diameter of the 21st bright fringe for a wavelength of 500 nm if the etalon plates are 5.00 mm apart and air is the medium between the plates. You may assume that light is incident on the etalon near normal incidence.

The light will be incident on the etalon at a constant angle θ if it travels along the generators of a cone of semi-angle θ. This means that the projected fringes will consist of concentric bright and dark rings; each ring is obtained by rotating the diagram about the optical axis.

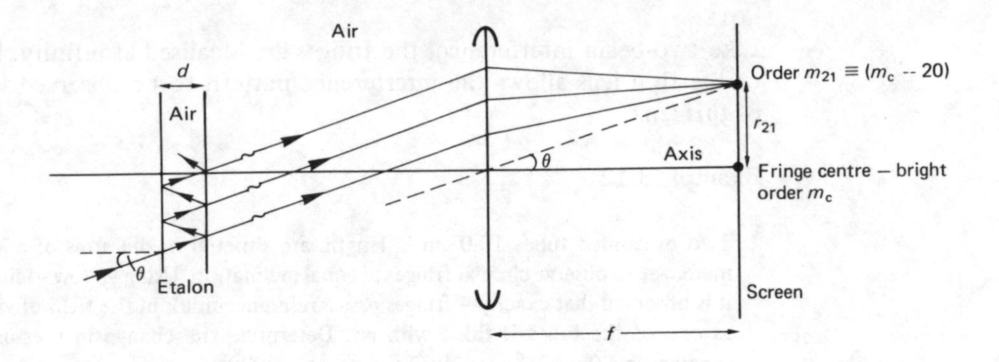

Data	Given:	$\lambda = 500$ nm
		$d = 5.00$ mm
		$f = 1.00$ m
	Unknown:	θ
		r_2
		m_2
		m_c
Relevant equations		$I_T = (I_T)_{max}/[1 + 4R \sin^2 (\delta/2)/(1-R)^2]$
		$2d \cos\theta = m\lambda$ for bright fringes

$$\cos\theta = m\lambda/2d \tag{1}$$

and

$$\tan\theta = r/f = D/2f \tag{2}$$

Therefore, because θ is small,

$$\cos\theta \simeq 1 - \theta^2/2; \tan\theta \simeq \theta$$

and (1) becomes

$$1 - D^2/8f^2 = m\lambda/2d$$

from which

$$D^2 = 8f^2 \ (1 - m\lambda/2d) \tag{3}$$

At the centre of the fringe pattern $\theta = 0$ and

$$1 = m_c\lambda/2d$$

from which

$$m_c = 2d/\lambda \tag{4}$$

Therefore

$$m = 2d/\lambda - 20 \tag{5}$$

Hence

$$\begin{aligned} D &= 2f\,[2\,(1 - (2d/\lambda - 20)\,\lambda/2d)]^{\frac{1}{2}} \\ &= 2f\,[2\,(1 - 1 + 10\lambda/d)]^{\frac{1}{2}} \\ &= 2f\,(20\lambda/d)^{\frac{1}{2}} \\ &= 4f\,(5\lambda/d)^{\frac{1}{2}} \\ &= 4 \times 1.00 \times (5 \times 5.0 \times 10^{-7}/5.00 \times 10^{-3})^{\frac{1}{2}}\ \text{m} \\ &= 4 \times 1.00 \times (5.0)^{\frac{1}{2}} \times 10^{-2}\ \text{m} \\ &= 8.9\ \text{cm} \end{aligned} \tag{6}$$

NOTE

Like two-beam interference, the fringes are localised at infinity. However, the use of the thin lens allows the interference pattern to be observed in the focal plane of this lens.

Example 3.12

> Two evacuated tubes 10.0 cm in length, are situated in the arms of a Michelson interferometer set to observe circular fringes of equal inclination. Using yellow sodium light (589.3 nm), it is observed that exactly 4 fringes pass a reference mark in the field of view of the telescope as one of the tubes is filled with air. Determine the change in pressure if the sensitivity constant is 3.0×10^{-4} atm^{-1}. (Take $n_{vac} = 1.0000$)

Circular fringes of equal inclination will be observed: (a) if the two mirrors at the extremities of the arms of the interferometer are at right angles, and (b) if an extended source is used to allow light to fall on the beam-splitter at more than one angle of incidence with roughly equal irradiance.

With the tubes evacuated, and for near normal incidence, the condition for bright fringes can be written

$$2d = m\lambda$$

if the beam splitter has a partially metallised surface. d is the difference between the length of the arms. Now, as one of the tubes is filled with air to a pressure P the optical path difference changes by an amount $2L(n_P - 1.000)$, where L is the tube length and n_P is the refractive index of air at pressure P. Fringes will be observed to move across the field of view. Let this number be N, which may be non-integral. Then the optical path difference is also $N\lambda$.

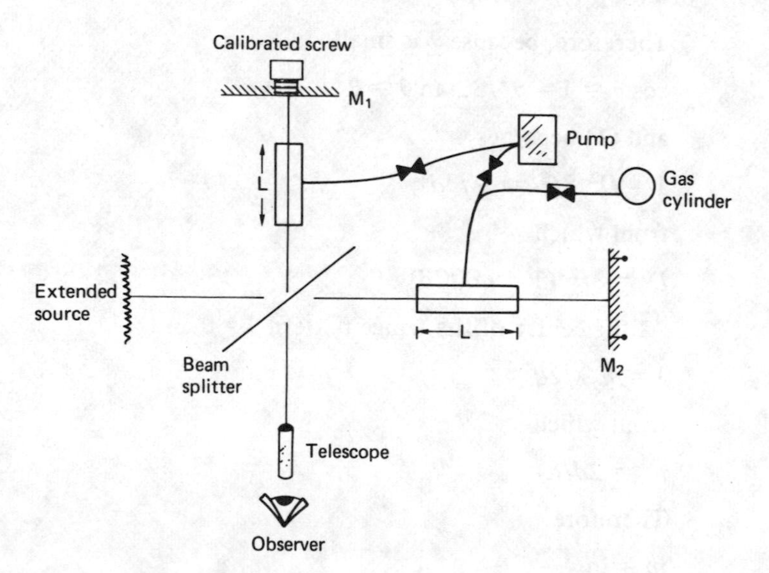

Data	*Given:*	$\lambda = 589.3$ nm
		$N = 4$
		sensitivity constant $= 3.0 \times 10^{-4}$ atm^{-1}
		$L = 10.0$ cm
		$n_{\text{vacuum}} = 1.000$
	Unknown:	d
		m
		pressure P
		n_P
Relevant equations		$2L\,(n_P - 1.000) = N\lambda$
		$n_P = 1.000 + kP$ (k = sensitivity constant)

We have

$$n_P = 1.000 + (N\lambda/2L) \tag{1}$$

and

$$P = (n_P - 1.000)/k$$
$$= N\lambda/2Lk \tag{2}$$
$$= 4 \times 5.893 \times 10^{-5}/2 \times 10.0 \times 3 \times 10^{-4} \text{ atm}$$
$$= 3.9 \times 10^{-2} \text{ atm}$$
$$= 3.9 \times 10^{3} \text{ Pa}$$

NOTES

(i) With a metallised beam splitter there is no phase change of π on reflection.

(ii) To an observer, the formation of the fringe system occurs in an identical manner to that in parallel-block interference.

(iii) M_1 can be moved longitudinally only and distances are read using a calibrated screw. M_2 is fixed except for rotational movement – with this, fringes of equal thickness may be observed.

(iv) Historically, the interferometer is important because it was used to demonstrate that the speed of light is independent of the Earth's motion – one of the postulates of The Special Theory of Relativity.

(v) The interferometer has many other applications, such as investigating the tidal effects of the Moon on the Earth and continental drift.

Example 3.13

A Michelson interferometer is used to investigate a spectral doublet with wavelengths λ and $\lambda + \Delta\lambda$. Fringes of equal inclination are used and discordances are observed to occur when the difference between the length of the arms of the interferometer attains certain values. Three such values are 0.16 mm, 0.44 mm and 0.74 mm. Determine the value of $\Delta\lambda$ if λ is 600 nm.

Each wavelength in a light source will give rise to its own fringe system. A concordance will occur if the two fringe systems coincide exactly and a discordance when the bright fringes of one system coincide with the dark fringes of the other. The field of view in the latter case is uniformly illuminated.

Data	*Given:*	$d_1 = 0.16$ mm
		$d_2 = 0.44$ mm
		$d_3 = 0.74$ mm
		$\lambda = 600$ nm
	Unknown:	m
		$\Delta\lambda$
Relevant equations		$2d = m\lambda$ (bright fringes)
		$= (m + B)\lambda$ (dark fringes; $B = \frac{1}{2}, \frac{3}{2}, \ldots, (2r - 1)/2; r =$ integer)

Bright fringes for λ and dark fringes for $\lambda + \Delta\lambda$ coincide when

$$m\lambda = (m + B)(\lambda + \Delta\lambda)$$
$$= m\lambda + m\Delta\lambda + B\lambda + B\Delta\lambda \quad (1)$$

or

$$m = -B\lambda/\Delta\lambda \quad \text{(neglecting } B\Delta\lambda \text{ compared with } B\lambda) \quad (2)$$

The negative sign need not concern us. It means that as λ increases m decreases, and vice versa.

Substituting for m in

$$2d = m\lambda$$

we have

$$2d = B\lambda^2/\Delta\lambda \quad (3)$$

Thus

$$\Delta\lambda = (r - \tfrac{1}{2})\lambda^2/2d_r \quad (4)$$

Each of the given values of d_r correspond to a particular value of r. A table of values can, therefore, be drawn up which gives values of $\Delta\lambda$ calculated using

$$\Delta\lambda = (r - \tfrac{1}{2})(6.00 \times 10^{-7})^2/2d_r$$

r	$d_r/10^{-4}$ m	$\Delta\lambda/10^{-10}$ m
1	1.6	5.6
2	4.4	6.1
3	7.4	6.1

The mean value of $\Delta\lambda = (5.9 \pm 0.2) \times 10^{-10}$ m

NOTES

(i) It is assumed that the two wavelengths are emitted with equal irradiance. If this is not the case then a 'true' discordance will not be obtained.

(ii) The analysis is correct near normal incidence when cos r can be replaced by 1.

(iii) The Michelson interferometer was originally used to investigate the structure of spectral lines although it has been superseded by the Fabry–Perot etalon.

(iv) This is an idealised experiment because it is assumed that the radiation emitted by the source consists of two discrete wavelengths. In actual fact, all spectral lines have a certain width, which implies that the radiation has a wavelength spread centred at λ_o and extending from $\lambda_o - \Delta\lambda$ to $\lambda_o + \Delta\lambda$. With such a spectral line the fringe visibility gradually decreases as d increases; the different fringe systems generated by the various wavelengths only truly coincide for d equal to zero, and never coincide exactly again as d is increased.

3.6 Questions

3.1

White light falls normally on a horizontal oil film floating on water. If the refractive indices of the oil and water are, respectively, 1.800 and 1.340 and the thickness of the oil film is 5.0×10^{-7} m find the number of wavelengths in the visible spectrum which reflect strongly.

3.2

A glass substrate (n = 1.600) is coated with a transparent thin film (n = 1.500). When white light is incident normally on the combination destructive interference occurs for wavelengths of 428.6 and 600 nm. Calculate the thickness of the transparent film.

3.3

The side section of a piece of plastic (n = 1.500) has the shape of a truncated wedge. When light of wavelength 600 nm is incident on the plastic at an angle of 30.0° it is observed that there are 20 fringes between its ends. Find the difference in thickness between the ends of the plastic.

3.4

Two pieces of metal, nominally of equal length, have ends which have been accurately polished so that they are flat and parallel. The metal pieces are placed 5.0 cm apart on an optical flat and a thin glass plate is placed over them. Using light of wavelength 600 nm, it is observed that 8 fringes occur per centimetre. Find the difference in the length of the metal pieces.

3.5

In a Young's slits experiment using light of wavelength 550 nm a transparent material of refractive index 1.583 and thickness 6.6×10^{-6} m is placed over one of the slits. Show that the effect of the material is to displace the fringe system so that the seventh bright fringe occupies the central point of the observing screen. (Ignore refraction at the mica.)

3.6

A Fabry-Perot etalon has plates 1.00 cm apart and is used in the transmission mode. If the coefficient of reflectance of the plates is 0.9 determine the minimum wavelength difference that can be resolved by the etalon on each side of the central wavelength of 500 nm.

3.7

In the Lloyd's mirror technique, a light beam from a pin-hole source travels directly to a screen, a distance D away, whilst another beam is specularly reflected from a glass plate. Prove that there is a dark fringe on the axis of the system, i.e. where a line drawn in the surface of the plate meets the screen normally. Show that the arrangement similates a Young's slits set-up. Determine the fringe separation for light of wavelength 500 nm if the pin-hole-to-screen separation is 1.00 m and the distance between the pin-hole and its virtual image is 5.0 mm.

3.8

The insertion of a thin transparent piece of mica into one arm of a Michelson interferometer produces a movement of 20 fringes. Determine the thickness of the mica if light of wavelength 589.3 nm is used. Take the refractive index of the mica to be 1.583.

3.9

Determine the visibility of the interference fringe system produced when monochromatic light is reflected at a glass block. Assume that you are operating near normal incidence and that the amplitude reflection coefficient is 20%.

3.10

Obtain an expression for the visibility of the fringe system produced by the interference of light transmitted by a glass block. Determine its value for an amplitude reflection coefficient of 0.2. Compare your value with that found from Question 3.9 and, thereby, convince yourself why the reflected system is always used in experiments on two-beam interference.

3.11

A glass block has its surfaces partially metallised. Interference fringes are produced when N transmitted waves superpose at infinity or in the second focal length of an objective lens. Following the method given in Example 2.6, obtain an expression for the resultant displacement. At each surface a fraction r of the amplitude is reflected and a fraction t is transmitted; absorption can be neglected.

3.12

Fringes of equal inclination are formed:

(a) when two beams reflected from the front and back surfaces of a parallel-sided block interfere;
(b) only when the angle of incidence at the front surface of a parallel-sided block is constant;
(c) by varying the thickness of the interference region but keeping the angle of incidence constant;
(d) when parallel beams are reflected from the front and back surfaces of a block of constant thickness and which satisfy the condition: $2nt \cos r = m\lambda$;
(e) none of these but . . .

3.13

Fringes of equal thickness are:

(a) always straight lines parallel to the line of intersection of the wedge plates;
(b) similar to height contours on a geographical map;
(c) only formed when the incident light is normal to the wedge;
(d) equally spaced straight lines for all values of the wedge angle;
(e) none of these but . . .

3.14

An anti-reflection coating consists of:

(a) magnesium fluoride deposited on glass to a thickness of $\lambda/4$;
(b) a thin film of refractive index greater than air but less than the substrate material and of optical thickness $\lambda/4$;
(c) a thin film of refractive index equal to the geometric mean of the refractive indices of the substrate and air and a thickness of $\lambda/4$;
(d) a thin film of refractive index equal to the geometric mean of the refractive indices of the substrate and the medium in which it is immersed and of optical thickness $(2m + 1)\lambda/4$, where m is integral or zero.
(e) none of these but . . .

3.15

A phase change of π occurs:

(a) always on reflection;
(b) when a wave is travelling in the less dense of two media and is reflected at the interface between them;
(c) when a wave is travelling in the denser of two media and is reflected at the interface between them;
(d) in the layers of a dielectric stack of optical thickness $\lambda/4$;
(e) none of these but . . .

3.16

The optical path difference between two waves:

(a) depends on the geometrical path difference for all wavelengths;
(b) depends on the phase angle of each wave only;
(c) is identical to the path difference in air;
(d) is the product of the geometrical path difference and the refractive index of the medium in which they are travelling;
(e) none of these but . . .

3.17

The Michelson interferometer:

(a) has two arms which lead to plane mirrors which must be at right angles to one another before fringes can be observed;
(b) can be likened to a Wheatstone bridge electrical circuit;
(c) must use an extended light source for high visibility fringes to be observed;
(d) is based on the principle of division of amplitude;
(e) none of these but . . .

3.18

The Twyman–Green interferometer:

(a) is a modification of the Michelson interferometer;
(b) must use an extended source of light;
(c) assesses the variation in the light paths across an optically transparent material by observing the number of fringes that are produced;
(d) assesses the degree of flatness of the ends of a ruby rod for a laser;
(e) none of these but . . .

3.19

In a Young's slits experiment the primary wave arriving at a slit and the secondary waves leaving it are:

(a) always in phase;
(b) out of phase by $\pi/2$;
(c) out of phase by π;
(d) figments of the imagination;
(e) none of these but . . .

3.20

The degree of flatness of a transparent plate can be assessed by placing the plate in contact with an optical flat and viewing wedge-type fringes. You will observe that:

(a) the fringes cover the whole of the field of view;
(b) the fringes are similar to Newton's rings but are confined to a small area in the field of view;
(c) the number of fringes are an indication of the size of a surface protrusion or pit;
(d) the fringes are distorted straight lines;
(e) none of these but . . .

3.21

The thickness of a glass block is important in two-beam interference because:

(a) the visibility of the fringe system decreases with thickness;
(b) the beams become less spatially-coherent;
(c) there is a phase change of π at the glass/air interface;
(d) the number of interference fringes increases with thickness;
(e) none of these but . . .

3.7 Answers to Questions

3.1 400 nm; 514 nm; 720 nm

There is one phase change of π at the upper surface of the oil. The condition for constructive interference is

$$2nd = (m + \tfrac{1}{2})\lambda$$

As there are two unknowns in this relation the technique to use is to try various values of m until λ lies between 350 nm and 700 nm.

3.2 500 nm

There is a phase change of π at the upper and lower surfaces of the film. The condition for destructive interference is

$$2nd = (m + \tfrac{1}{2})\lambda$$

For each value of λ try different values of m until you find

$$(m_1 + \tfrac{1}{2})\lambda_1 = (m_2 + \tfrac{1}{2})\lambda_2 = 1800 \text{ nm}$$

3.3 4.6 μm

Write down the interference relations at both ends of the truncated wedge, where the thicknesses are d_1 and d_2. These are:

$$2nd_1 \cos\theta = m\lambda$$
$$2nd_2 \cos\theta = (m + 20)\lambda$$

Subtract to determine $(d_2 - d_1)$.

3.4 11.5 μm

There are no phase changes on reflection from a metal surface. In going from one fringe to the next the wedge thickness changes by $\lambda/2$.

3.5 For a transparent material of refractive index n and thickness x show that at the point on the screen where the axis touches it the optical path difference between light from the sources is $(n - 1)x$. Then

$$(n - 1)x = \Delta m\, \lambda$$

where Δm is the shift in the order of interference. It is interesting to note that if x was sufficiently large no fringes would be observed (see Section 3.4). This means that one beam is delayed by a distance at least as large as the (temporal) coherence length of the wave train.

3.6 0.4 pm

The chromatic resolving power is 1.49 $m(4R/(1 - R)^2)^{\frac{1}{2}}$. The only unknown is m which can be found from

$$2d = m\lambda$$

3.7 1.0×10^{-4} m

This is an example of two-beam interference. If the optical path lengths of the direct beam and reflected beam are, respectively, L_1 and L_2, then the phase difference δ at some point on the screen will be

$$(2\pi/\lambda)(L_1 - L_2) + \pi$$

The π term appears because of the phase change on reflection at the plate. Hence the irradiance distribution of the fringes is the usual cosine² dependence, i.e.

$I = I_o \cos^2 \delta/2$

$= I_o \cos^2 \left[(\pi/\lambda)(L_1 - L_2) + \frac{\pi}{2}\right]$

$= I_o \sin^2 [(\pi/\lambda)(L_1 - L_2)]$

If distances along the screen are given the parameter x, then from the similarity between this arrangment and the Young's slits set-up $(L_1 - L_2)$ is identical with $2zx/D$. Hence at $x = 0$, $I = 0$. The fringe separation is given by $\lambda D/2z$.

3.8 1.000×10^{-5} m

If the thickness of the mica is d, then the change in the optical path difference between the arms of the interferometer is $d(n - 1)$. A shift of 1 fringe corresponds to a change in the optical path difference of $\lambda/2$. Hence

$d(n - 1) = 20\lambda/2$

3.9 800/881

Use the following diagram to assist with this question and Question 3.10, where A is the amplitude of the incident light.

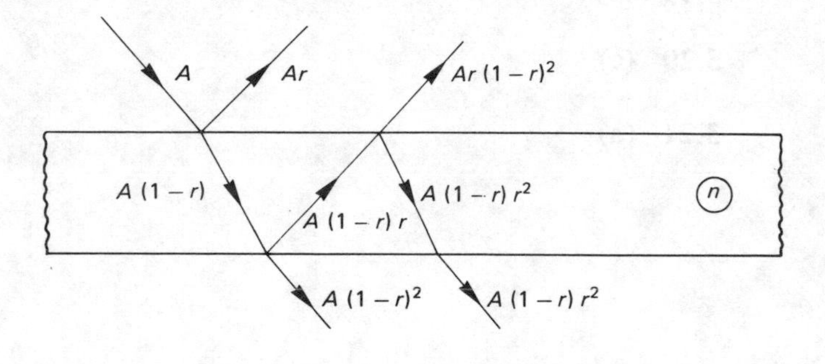

Then

$I_{max} = k\,[Ar\,(1 + (1 - r)^2]^2$; $I_{min} = k\,[Ar\,(1 - (1 - r)^2]^2$

Substitute into the fringe visibility expression. k is a constant of proportionality.

3.10 40/401

From the diagram in Question 3.9:

$I_{max} = k\,[a\,(1 - r)(1 - r + r^2)]^2$

and

$I_{min} = k\,[a\,(1 - r)(1 - r - r^2)]^2$

3.11 $Y_{res} = MAt^2 \sin(2\pi ft + \phi)$

where

$M \exp(j\phi) = 1/[1 - r^2 \exp(-j\delta)]$

The resultant irradiance obtained from

$I_{res} = kM^2A^2t^4$

is called the Airy irradiance distribution (see standard textbooks).

3.12 (e)

The angle of incidence must vary to give a family of fringes but not all values are acceptable (see Section 3.2(a)).

3.13 (b)

3.14 (d)

3.15 (e)

(b) is correct if the more optically-dense medium is a dielectric.

3.16 (e)

It is conceivable that the waves are travelling in different media, in which case the optical path of each wave must be calculated separately. (This must include any phase changes of π.)

3.17 (b); (c); (d)

3.18 (a); (c); (d)

3.19 (b)

3.20 (c)

3.21 (a)

4 Diffraction

4.1 Definition

Diffraction can be used to describe the departure of radiation from its rectilinear path, such as may occur when it meets a slit or a circular aperture. The important parameter to be determined is the ratio wavelength λ: characteristic aperture size a; for a slit, a would be its width and for a circular aperture it would be the diameter. If λ/a is $\ll 1$, diffraction can be ignored and the transmitted wavefront propagates in the same direction as the incident wavefront. However, if λ/a is $\gg 1$ the energy of the incident wave is uniformly distributed in all directions (see Fig. 4.1).

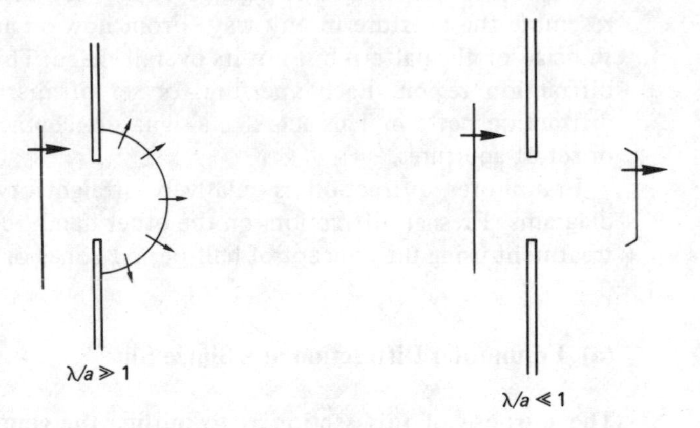

Figure 4.1

4.2 Huyghens' Principle

Huyghens' principle of *secondary sources* is used to study the propagation of radiation in *free space*, i.e. in the absence of apertures and obstacles. Each point on the wavefront acts as a *secondary source* emitting light (secondary wavelets) of the same frequency as the primary source in all directions. The new position of the wavefront, i.e. after a time interval t, is given by the tangent surface to all the secondary wavelets in the forward direction. Thus a circular wavefront propagates as a circular wavefront. This is a useful principle but it has two main deficiencies: (i) no account is taken of those portions of the secondary wavelets which are not common to the tangent surface; (ii) the wavelength of the light is ignored.

4.3 Huyghens–Fresnel Principle

Fresnel extended Huyghens' principle in an attempt to overcome these objections. The Huyghens–Fresnel principle must be used when there are apertures and obstacles in the way of the propagating wavefront. It says that at every instant of time each point on the *unobstructed* portion of a wavefront acts as a source of spherical secondary wavelets. The resultant displacement on the other side of the aperture or obstacle must be found by applying the Principle of Superposition to the waves emitted by these secondary sources. Although the deficiencies of the original Huyghens' model are largely overcome, an exact description of the propagation of light requires a highly mathematical treatment (see Kirchhoff and the electromagnetic theory of light).

Note that with apertures in the form of slits, each slit is divided into an infinite number of elementary slits which act as secondary sources. As the amplitude of the light incident on the slit as a whole is finite, this means that the amplitude of the light emitted by each secondary source is infinitesimally small.

4.4 Types of Diffraction

Imagine that a screen has a single small aperture in it which is illuminated by plane waves. If a second screen is placed close to the first one, an image of the aperture will be observed on it. As the second screen is moved away the image quality falls off and fringes are observed near the boundary of the original image. These fringes are produced by *Fresnel* or *near-field* diffraction. Further movement causes the fringes to change to a more diffused pattern which does not resemble the aperture in any way. From now on any further movement increases the size of the pattern but not its overall shape. This is the *Fraunhofer* or *far-field* diffraction region. Each aperture, or set of apertures, has its own Fraunhofer diffraction pattern. This acts like a signature characterising the particular aperture or set of apertures.

Fraunhofer diffraction is relatively straightforward to analyse using phasor diagrams. Fresnel diffraction, on the other hand, requires a rigorous mathematical treatment using the concept of half-period zones or strips.

(a) Fraunhofer Diffraction at a Single Slit

The purpose of this section is to outline the general approach that needs to be adopted in order to obtain an expression for the resultant amplitude at infinity under Fraunhofer diffraction conditions.

The slit has a width a and light is incident normally. Following the Huyghens–Fresnel principle every elementary slit on the unobstructed portion of the wavefront in the plane of the slit acts as a secondary source. We shall imagine that an observer is at infinity looking back at the slit in some direction which makes an angle θ with the axis, as in Fig. 4.2. Then, if the separation between neighbouring secondary sources is Δa, we can write the phase difference between waves from them as

$$\delta = (2\pi/\lambda)\,(\Delta a \sin\theta) \qquad (4.1)$$

Δa is infinitesimally small. Each secondary source is coherent because the incident wave arrives at every point in the slit at the same instant of time.

Let the displacement of the wave from the secondary source at C at infinity at time t be y, then

$$y = \Delta A \sin\,[(2\pi/\lambda)\,(x - vt)] \qquad (4.2)$$

Similarly, the wave from the secondary source at C_1 at time t has a displacement y_1, given by

$$y_1 = \Delta A \sin\,[(2\pi/\lambda)\,(x_1 - vt)] \qquad (4.3)$$

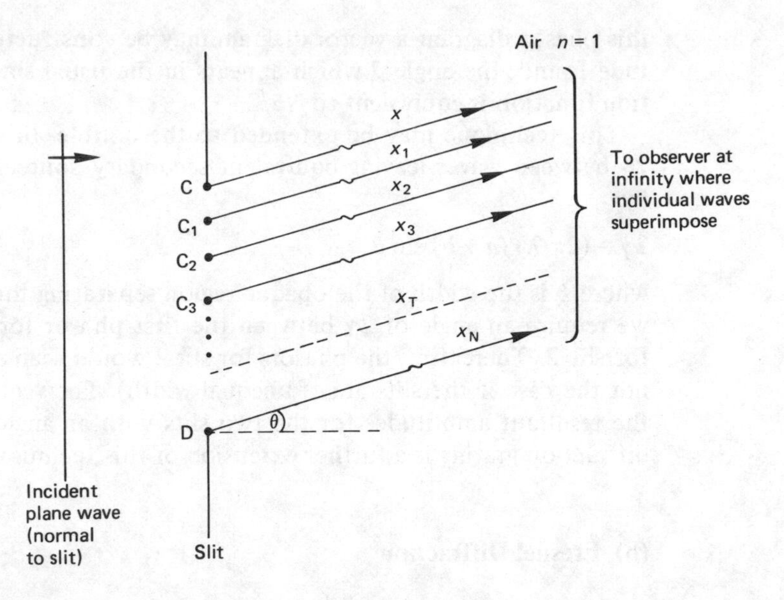

Figure 4.2

and so on for all the other secondary sources. ΔA is the amplitude of the light emitted by each secondary source. If there are N secondary sources then it is not unreasonable to write

$$N\Delta A \approx A \tag{4.4}$$

where A is the amplitude of the incident wave. N, of course, will tend to infinity.

We know from Section 2.4 that the phase difference between the two waves at infinity is

$$\delta = \psi_1 - \psi = (2\pi/\lambda)(x_1 - x) = \ldots = \psi_N - \psi_{N-1} \tag{4.5}$$

Hence the phasor diagram for superposing the wave motions consists of an infinite number of phasors, each separated by an angle δ, as depicted in Fig. 4.3. From

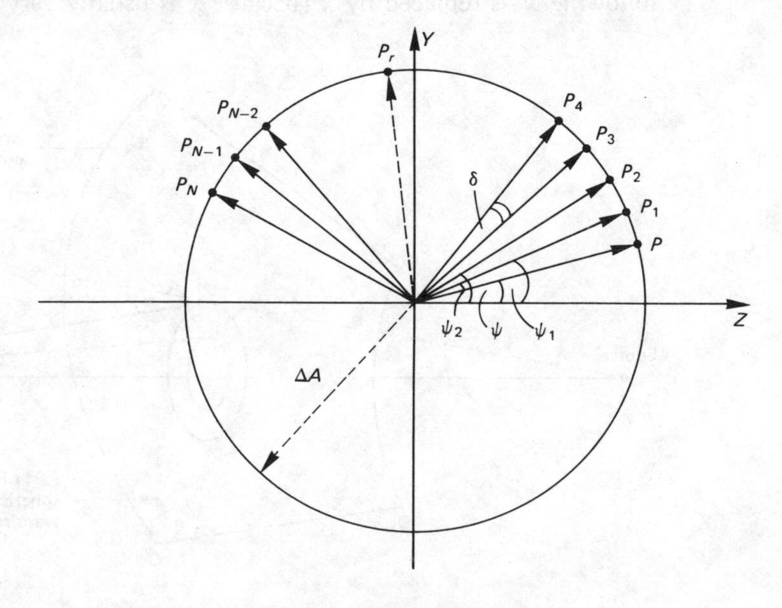

Figure 4.3

this phasor diagram a vector diagram may be constructed and the resultant amplitude found; the angle β which appears in the usual single slit irradiance distribution function is equivalent to $N\delta/2$.

This technique may be extended to the double-slit case. The phase difference 2γ between waves leaving equivalent secondary sources in the two slits is defined by

$$2\gamma = (2\pi/\lambda)\,(a + b)\sin\theta \tag{4.6}$$

where b is the width of the opaque region separating the slits. Then, using Fig. 4.3 we require an angle of 2γ between the first phasor for slit 1 and the first phasor for slit 2. Thereafter, the phasors for slit 2 would scan an equal sector area (this is not the case if the slits are of unequal width). The vector diagram is formed from the resultant amplitudes for the two slits with an angle of 2γ between them. The diffraction grating is a further extension of this technique.

(b) Fresnel Diffraction

The procedure outlined in Section 4.4(a) cannot be applied when the observer is close to the diffracting element because the phase difference between waves from adjacent secondary sources is no longer constant. In this case the wavefront is divided into a number of regions called *half-period* zones. Then the periphery of zone r is an optical distance of $\lambda/2$ further away from the observer than the periphery of zone $(r - 1)$. This means that waves from these peripheral regions arrive at the observer with a phase difference of π.

Consider Fig. 4.4 in which a source S generates the spherical wavefront CDEB. At the instant being considered the wavefront is a distance x_o from S and the observer is a distance x from the centre O of the wavefront. Two half-period zones, only, are shown for the sake of clarity. A detailed study shows that the zones are roughly equal in area, so that they contain the same number of secondary sources. Further,

$$m\lambda/2 = a_m^2\ [(x_o + \bar{x})/2x_o\bar{x}] \tag{4.7}$$

Here $\bar{x}$ is the arithmetic mean of x and $(x + m\lambda/2)$, although in the examples following $\bar{x}$ is replaced by x (because x is usually very much larger than $m\lambda/2$).

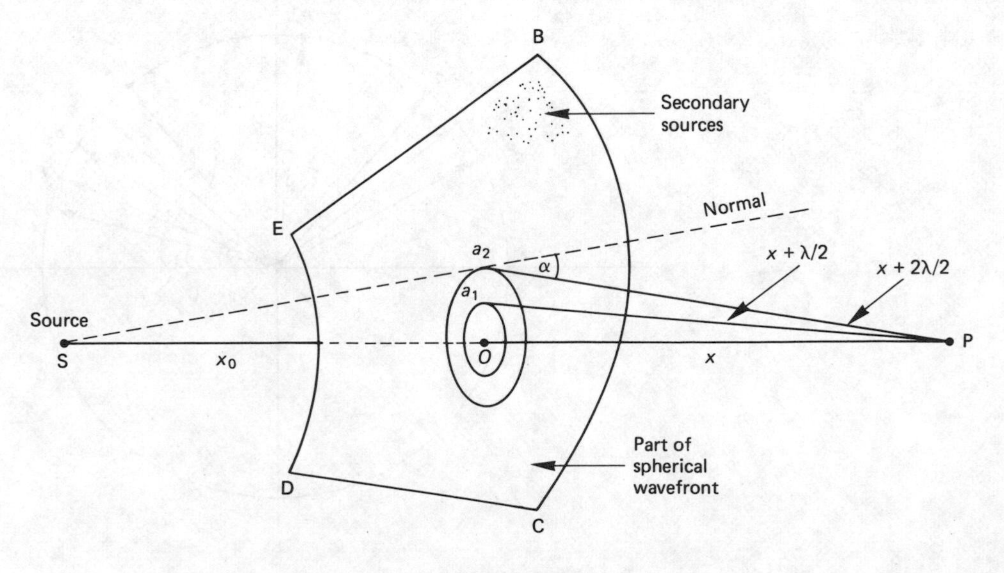

Figure 4.4

The light waves from the various secondary sources are directed downwards to the axial observation point which means that the angle α between this direction and the forward normal gradually increases as we move further away from the axis. It takes all values from zero to π (for a completely unobstructed wavefront). This angle is called the *obliquity* factor.

The resultant amplitude A_P at the observing point P may be written

$$A_P = A_1 - A_2 + A_3 - A_4 + \ldots + (-1)^{m-1} A_m \tag{4.8}$$

where the alternating nature of the signs takes account of the phase difference of π between adjacent half-period zones and A_r is the amplitude of light from the rth zone. For an unobstructed wavefront, A_P can be shown to be given by

$$A_P = A_1/2 \pm A_m/2 \tag{4.9}$$

The positive sign is used if there are an odd number of zones and the negative sign with an even number.

(i) *The Cornu Spiral*

With diffraction at straight edges and rectangular slits, the emerging wavefront is cylindrical and the half-period zones are strips. If each strip is divided into an infinite number of sub-strips then the vector diagram at P is a continuous curve called the *Cornu Spiral*. It is described by the parametric equations:

$$y = \int_0^{\nu} \cos^2 (\pi \nu^2/2)\, d\nu$$

$$z = \int_0^{\nu} \sin^2 (\pi \nu^2/2)\, d\nu \tag{4.10}$$

These are the Fresnel Integrals and a table will be found in all optics textbooks that deal with this subject. ν is the length of an arc of the spiral measured from the origin. Figure 4.5 illustrates the Cornu Spiral for four half-period strips on each side of the origin. ν can be defined in terms of the phase difference δ at P between waves from different half-period zones. The relation between them is

$$\delta = (\pi/2)\, \nu^2 \tag{4.11}$$

Using the expression (4.7) given in Section 4.4(b) we find that ν may be written as

$$\nu = a_m \; [2\,(x_o + x)/x_o x\lambda]^{\frac{1}{2}} \tag{4.12}$$

It is appropriate in many examples to erect axes (Y- and Z-) with the origin at the centre of the diffracting body so that a_m, in the above expression, is a generalised coordinate.

Let us now consider a single slit, as before, with the observing point P on the axis of the system. Imagine for the sake of argument that two half-period zones are 'seen', one above and one below the axis. Thus a_m will take two values, one positive and one negative, from which two values of ν can be determined. Call these $-\nu_1$ and $+\nu_1$. Now measure a distance ν_1 from the origin *along* the two halves of the Cornu Spiral. Draw a line between these points and measure its length (see line 1 in Fig. 4.5). This is the resultant amplitude at P. Now increase the width of the slit until four half-period strips are 'seen'. Obtain the corresponding values of ν, $\pm \nu_2$, say, and measure along the arcs, as before. Join the end points with a straight line and obtain line 2 in Fig. 4.5. Repeat this exercise and you will notice that the resultant amplitude at P varies in an oscillatory way until, ultimately, when the slit is infinitely wide the resultant amplitude is equivalent to

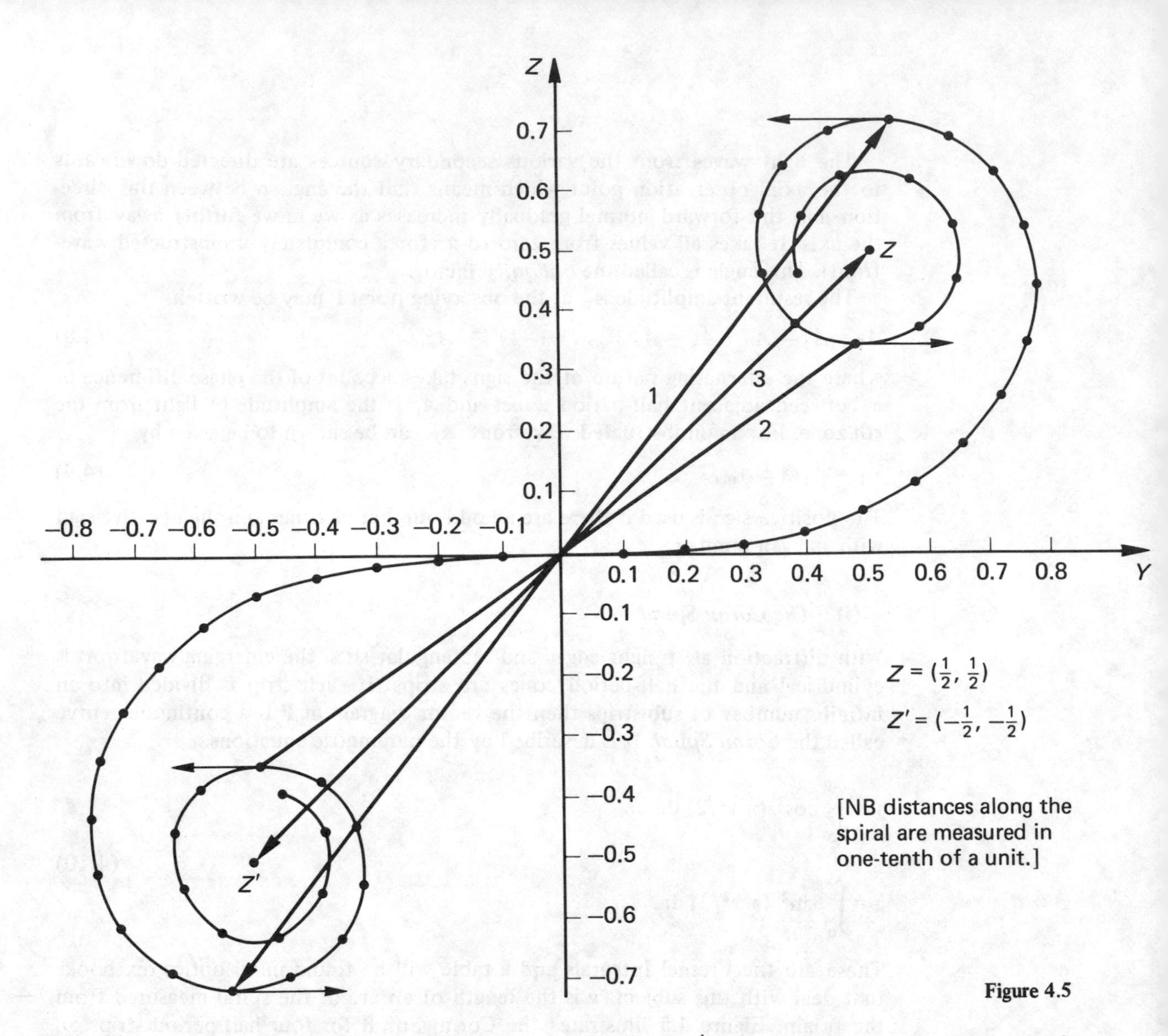

Figure 4.5

line 3 connecting Z and Z' (equal to $1/2^{\frac{1}{2}}$ units). The latter, then, is the resultant amplitude at P for an unobstructed wavefront.

If the observation point is below the axis the procedure is to translate the slit until this new point is again on the central axis. In other words, if the new observation point is a distance z below the axis we need to find values of v corresponding to $-(a_m + z)$ and $(a_m - z)$. The region of the Cornu Spiral now under consideration slides towards the lower branch of the spiral.

A similar technique can be applied to the case of diffraction at a straight edge (or knife edge).

4.5 Worked Examples

Example 4.1

Sodium yellow light of mean wavelength 589.3 nm is diffracted by a long narrow slit placed in front of a thin converging lens of focal length 50.0 cm. It is observed that the first minima in the Fraunhofer diffraction pattern are 5.00 mm apart. Calculate the width of the slit.

To determine what degree of diffraction occurs at the slit the ratio wavelength: slit width needs to be investigated; diffraction will be significant if this ratio is very much greater than 1. As a consequence it follows that the overall irradiance of the diffraction pattern decreases.

The presence of the thin converging lens does not alter the diffraction process for parallel light will still leave the slit to produce a secondary maximum, say, at the same angle of diffraction as in the absence of the lens. The function of the lens is to bring the diffraction pattern closer to the slit so that measurements can be conveniently made along its secondary focal plane. The position of the diffraction minima in the secondary focal plane of the lens can be determined by tracing a line through the optical centre of the lens parallel to the diffracted beam and locating where it intersects the focal plane. This construction is only strictly accurate for a thin lens because any lateral displacement of the central ray can be ignored.

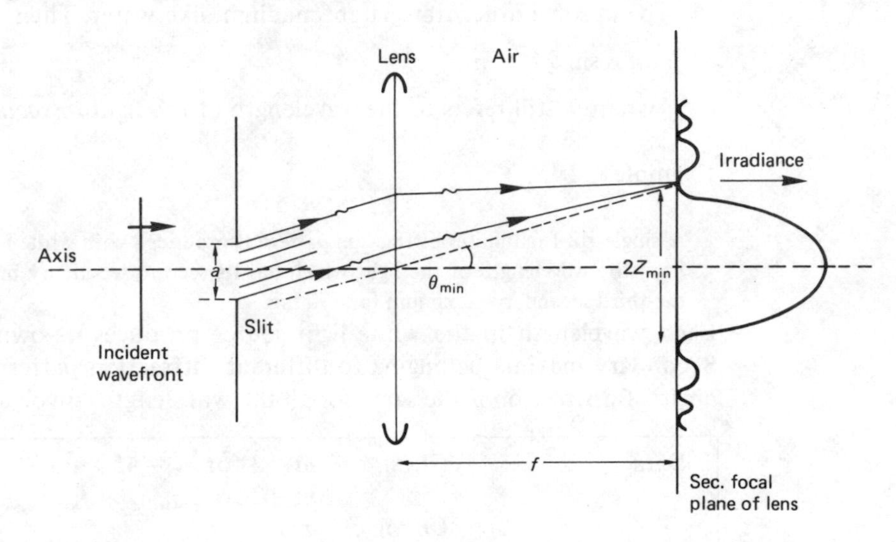

Data	*Given:*	$\lambda = 589.3$ nm
		$f = 50.0$ cm
		$2z_{min} = 5.00$ mm
	Unknown:	a
		irradiance distribution
		θ_{min}
Relevant equations		$I = I_o (\sin \beta/\beta)^2$; $\beta = (\pi a/\lambda) \sin \theta$
		$a \sin \theta = \pm m\lambda$
		$\tan \theta = z/f$

For small angles,

$\sin \theta \simeq \theta$ (radians)

At the first minima ($m = 1$)

$$\theta = \pm \lambda/a \tag{1}$$

but is also given by

$$\theta = z_{min}/f \tag{2}$$

So

$\lambda/a = z_{min}/f$ (omitting the ± signs)

from which

$$a = \lambda f/z_{min} \tag{3}$$
$$= 5.893 \times 10^{-5} \times 50.0/2.50 \times 10^{-1} \text{ cm}$$
$$= 1.2 \times 10^{-2} \text{ cm}$$

NOTES

(i) The incident light is assumed to be normal to the plane of the slit, with the result that the centre of the diffraction pattern lies on the optical axis of the system. With non-normal incidence there will be an additional optical path difference of $a \sin i$ to take into consideration in any analysis, and the centre of the pattern will be displaced from the axis.

(ii) The ± signs are not required in the calculation. The plus sign is used with angles of diffraction lying above the axis (as in the diagram) and the negative sign with those below it.

(iii) The usual experimental set-up is in air but there is no reason why it cannot be in some other transparent medium, like water. Then

$$an_w \sin \theta = \pm m\lambda$$

where λ still refers to the wavelength of the light *in vacuo* or in air.

Example 4.2

A single slit Fraunhofer diffraction pattern is produced with white light at normal incidence. Find the wavelength of the light which has its second secondary maximum coinciding with the third secondary maximum for 450 nm.

Each wavelength in the white light source produces its own diffraction pattern. Secondary maxima belonging to different diffraction patterns will coincide if the angle of diffraction is the same for all the wavelengths involved.

Data	*Given:*	$m = 3$ for $\lambda_k = 450$ nm
		$m = 2$ for λ_{un}
	Unkown:	a
		θ
		λ_{un}
		I/I_o
Relevant equations		$I = I_o (\sin \beta/\beta)^2$
		$\tan \beta = \beta$
		$\beta_2 = \pm 2.46\pi$ for $m = 2$ maximum
		$\beta_3 = \pm 3.47\pi$ for $m = 3$ maximum

We have

$$\beta_2 = \pi a \sin \theta/\lambda_{un} = 2.46\pi \text{ (taking + sign only)}$$

and

$$\beta_3 = \pi a \sin \theta/\lambda_k = 3.47\pi$$

Dividing, we find that

$$\lambda_k/\lambda_{un} = 2.46/3.47$$

and

$$\begin{aligned}\lambda_{un} &= 3.47\lambda_k/2.46\\ &= 3.47 \times 450/2.46\\ &= 635 \text{ nm}\end{aligned}$$

NOTE

The equation

$$\tan \beta = \beta$$

is a transcendental equation. It does not have a simple solution. Such equations must be solved numerically using mathematical techniques that gradually 'home-in' on the solution. Because β_2 does not equal 2.5π exactly the $m = 2$ secondary maximum does not lie midway between the $m = 1$ and $m = 3$ maxima. This statement can be obviously generalised to include all secondary maxima.

Example 4.3

Two long parallel slits are 0.40 mm wide and have their centres 2.00 mm apart. Which interference maxima would you expect to be missing from the diffraction pattern?

Certain interference maxima will be missing from the diffraction spectrum if the mark-space ratio is integral, i.e.the ratio of the centre–centre distance to slit width. The fact that theory suggests that an interference maximum should occur for certain angular positions on a screen is insufficient by itself to justify its presence. For, if the diffraction term is zero at these angular positions then the interference term will be suppressed. Hence the origin of the phrase 'missing orders'. If the mark-space ratio is equal to an integer p, then the pth, $2p$th, $3p$th, etc., maxima will be missing. This is a general rule which applies equally well to the diffraction grating.

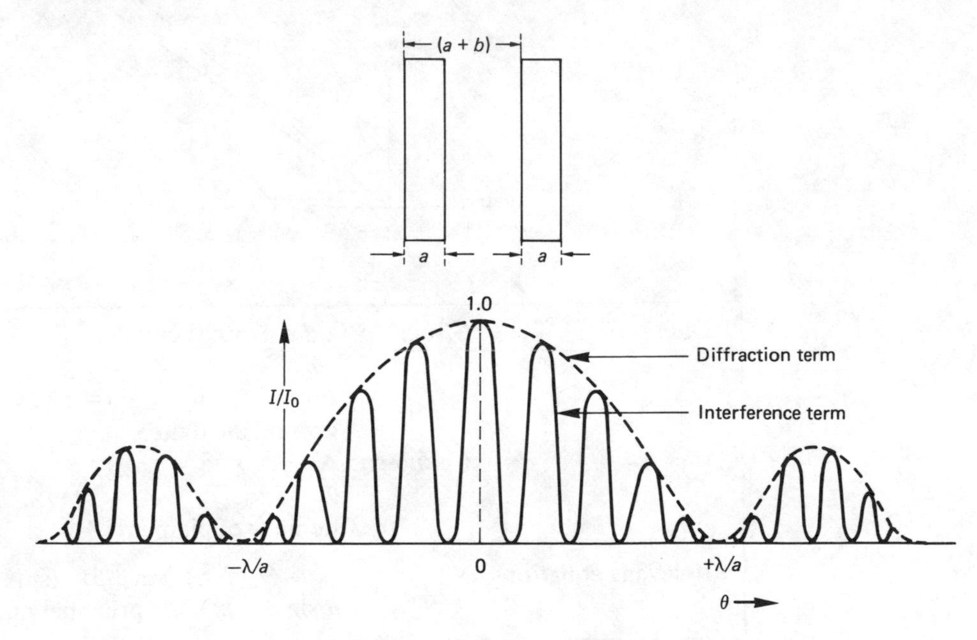

Data	*Given:*	$a = 0.40$ mm
		$(a + b) = 2.00$ mm
	Unknown:	I/I_o (except for $\theta = 0$)
		θ
		p
Relevant equations		$(a + b)/a = p$
		$I_2 = I_o\,(\sin\beta/\beta)^2\,\cos^2\gamma;\ \beta = \pi a \sin\theta/\lambda$ and $\gamma = \pi\,(a + b)\sin\theta/\lambda$

Now

$(a + b)/a = 2.00/0.40 = 5$

As p is integral the condition for suppression of certain interference maxima is satisfied.

The missing orders are the 5th, 10th, 15th, . . ., etc.

Example 4.4

The angular separation between the sodium D-lines in the second order spectrum produced by a plane transmission diffraction grating is 2.5 minutes of arc for normal incidence. If the grating has 5000 diffracting elements per centimetre, determine the wavelength separation between the D-lines.

Each wavelength in the light source produces its own interference pattern, modulated by diffraction, on transmission through the grating. The angular dispersion of the grating is an important parameter; the larger the angular dispersion the larger the angular separation in any given order m. The angular dispersion is not a constant, however, but depends directly on m and indirectly on the grating interval and the cosine of the angle of diffraction.

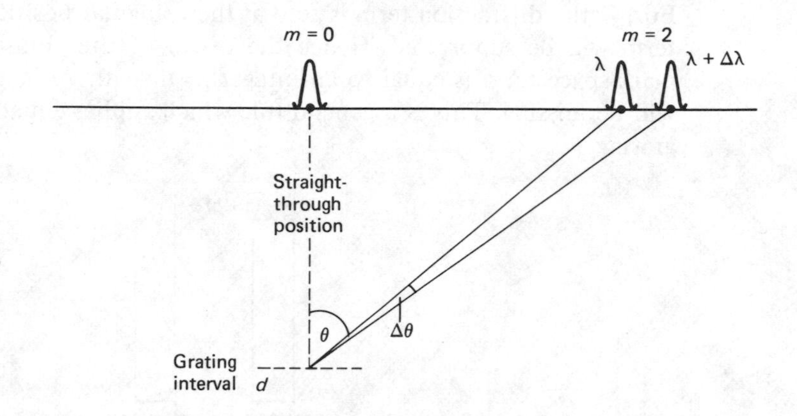

Data	Given:	$d = 1/5000$ cm
		$m = 2$
		$\Delta\theta = 2.5$ min of arc $= \pi/(24 \times 180)$ radians
		normal incidence
	Unknown:	$\Delta\lambda$
		θ
		number of elements N
Relevant equations		$I_N = (I_o/N^2)(\sin\beta/\beta)^2 (\sin N\gamma/\sin\gamma)^2$
		$d \sin\theta = m\lambda$ for principal maxima

The conditions for a principal maximum are:

$$d \sin\theta = m\lambda \tag{1}$$

$$d \sin(\theta + \Delta\theta) = m(\lambda + \Delta\lambda) \tag{2}$$

Expand (2) to give

$$d[\sin\theta \cos\Delta\theta + \cos\theta \sin\Delta\theta] = m(\lambda + \Delta\lambda) \tag{3}$$

Now use the small angle approximations:

$\sin\Delta\theta \to \Delta\theta$ (radians); $\cos\Delta\theta \to 1$

Thus (3) becomes

$$d(\sin\theta + \cos\theta . \Delta\theta) = m(\lambda + \Delta\lambda) \tag{4}$$

Subtract (1) from (4) to give

$$d \cos\theta . \Delta\theta = m\Delta\lambda$$

when

$$\Delta\lambda = d \cos \theta . (\Delta\theta/m)$$

Using

$$\cos \theta = (1 - \sin^2 \theta)^{\frac{1}{2}} = [1 - (m\lambda/d)^2]^{\frac{1}{2}}$$

we arrive at the final result, viz.

$$\Delta\lambda = (d/m) [1 - (m\lambda/d)^2]^{\frac{1}{2}} . \Delta\theta \qquad (5)$$

As λ appears in (5) there is need to assume a value for the wavelength of one of the sodium D-lines. We shall take λ to be 5.890×10^{-5} cm. Then

$$\Delta\lambda = (2.0 \times 10^{-4}/2) [1 - (2 \times 5.890 \times 10^{-5}/2.0 \times 10^{-4})^2]^{\frac{1}{2}} \times \pi/(24 \times 180)$$
$$= 5.6 \times 10^{-8} \text{ cm or } 0.56 \text{ nm}$$

NOTES

(i) The grating interval is also referred to as the grating element.

(ii) It was necessary to assume a value for the wavelength of one of the D-lines. The smaller of the two values was taken in order to agree with the diagram and equations (1) and (2). If θ is defined to be the angle of diffraction for the larger wavelength then

$$\begin{cases} d \sin \theta = m\lambda \\ d \sin (\theta - \Delta\theta) = m (\lambda - \Delta\lambda) \end{cases}$$

λ equal to 5.896×10^{-5} cm would now be used in the substitution.

(iii) The analysis used here applies to a 'square-wave' grating, i.e. one in which there is a sharp cut-off between the opaque and transparent regions: there is another type of transmission grating called the cosinusoidal grating (see Example 4.9).

Example 4.5

Light consisting of two wavelengths, 500 and 520 nm, is incident normally on a plane transmission diffraction grating. The grating element is 1.0×10^{-6} m, and a thin lens of focal length 2.00 m is used to observe the spectrum on a screen. Find the linear separation of the two first-order principal maxima.

The Fraunhofer diffraction pattern is observed in the secondary focal plane of the thin lens. The same technique as that used in Examples 3.7, 3.11 and 4.1 allows the principal maxima to be located.

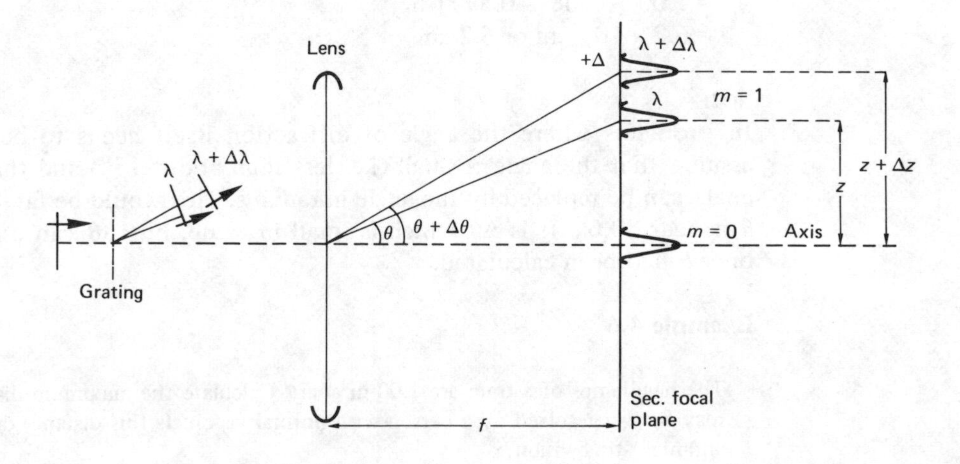

Data	*Given:*	$\lambda = 500$ nm
		$\lambda + \Delta\lambda = 520$ nm
		$d = 1.0 \times 10^{-6}$ m
		$f = 2.00$ m
		$m = 1$
		angle of incidence = 0°
	Unknown:	θ
		$\theta + \Delta\theta$
		z
		$z + \Delta z$
Relevant equations		$I_N = I_o\ (\sin\beta/\beta)^2\ (\sin N\gamma/\sin\gamma)^2$
		$d \sin\theta = m\lambda$
		$\tan\theta = z/f$

$\tan\theta = z/f$ for wavelength λ

and

$\tan(\theta + \Delta\theta) = (z + \Delta z)/f$ for wavelength $\lambda + \Delta\lambda$

Subtraction gives

$$\Delta z = f\,[\tan(\theta + \Delta\theta) - \tan\theta] \quad (1)$$

As

$\sin\theta = m\lambda/d$

or

$\theta = \sin^{-1}(m\lambda/d)$

we now have

$$\tan\theta = \tan[\sin^{-1}(m\lambda/d)] \quad (2)$$

and

$$\tan(\theta + \Delta\theta) = \tan[\sin^{-1}(m(\lambda + \Delta\lambda)/d)] \quad (3)$$

Hence substituting (2) and (3) into (1) allows Δz to be expressed as

$$\begin{aligned}\Delta z &= f\,[\tan[\sin^{-1}(m(\lambda + \Delta\lambda)/d)] - \tan[\sin^{-1}(m\lambda/d)]] \quad (4)\\ &= 2.00\,[\tan[\sin^{-1}(1 \times 5.20 \times 10^{-7}/1.0 \times 10^{-6})] - \\ &\qquad \tan[\sin^{-1}(1 \times 5.00 \times 10^{-7}/1.0 \times 10^{-6})]]\ \text{m}\\ &= 2.00\,[\tan[\sin^{-1} 0.52] - \tan[\sin^{-1} 0.50]]\ \text{m}\\ &= 2.00\,[0.608 - 0.577]\ \text{m}\\ &= 6.2 \times 10^{-2}\ \text{m or } 6.2\ \text{cm}\end{aligned}$$

NOTE

In problems where the angle of diffraction itself needs to be calculated never assume that the angle is small (i.e. less than about 15°) and that the sine of the angle can be replaced by the angle in radians. This would be fatal because, as here, θ is near 30.0°. It is $\Delta\theta$ which is small in value, but $\Delta\theta$ can only be determined once θ had been calculated.

Example 4.6

The headlamps of a tram are 1.00 m apart. Calculate the maximum distance at which they may be *just* resolved by a person with normal vision. Is this distance different for day-time and night-time vision?

This example is concerned with an observation which most of us have probably experienced. It involves Rayleigh's Criterion, which states that the diffraction images produced by two incoherent light sources will be just resolved when the principal maximum of one diffraction pattern occurs at the same angle of diffraction as the first minimum of the other. Each diffraction image is called an Airy disc. The disc consists of a series of bright and dark rings. With distant sources, for which the angle of diffraction θ is less than $1.22\lambda/D$, where D is the diameter of the eye's circular pupil, the Airy discs overlap. However, as the sources approach the observer the Airy discs begin to separate when θ is equal to $1.22\lambda/D$.

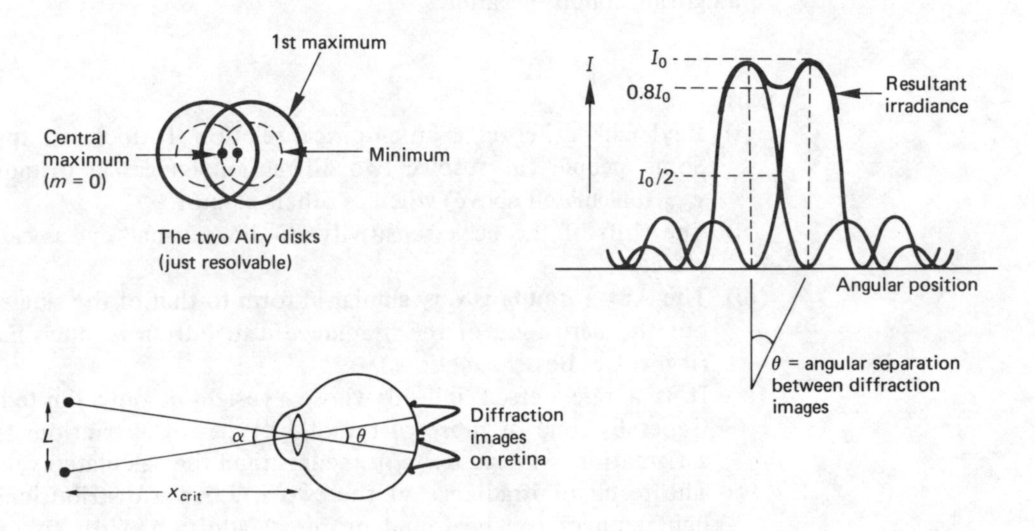

Data	*Given:*	$L = 1.00$ m
	Unknown:	λ
		α (and θ)
		D
		x_{crit}
		Irradiance distribution
Relevant equations		$\theta = 1.22\lambda/D$ $\alpha = L/x_{crit}$ assuming small angle approximation

The Airy discs are just resolvable when

$$\alpha = 1.22\lambda/D \quad (1)$$

i.e.

$$L/x_{crit} = 1.22\lambda/D$$

from which

$$x_{crit} = LD/1.22\lambda \quad (2)$$

As D and λ are unknown we have to make sensible estimates of their value.

In day-time, D is about 1–2 mm whereas at night it is about 8 mm. Thus the second part of the question can be answered immediately.

Further, as we are told in the question that the observer has normal vision, i.e. he/she does not suffer from aberrations of any kind, we can use the fact that the peak sensitivity of the eye occurs for λ equal to 505 nm under twilight conditions and 550 nm in daytime.

At twilight, with D about 6 mm and λ equal to 505 nm,

$$x_{crit} = 1.00 \times 8 \times 10^{-3}/1.22 \times 5.05 \times 10^{-7} \text{ m}$$
$$= 1.3 \times 10^4 \text{ m or } 13 \text{ km}$$

In daytime, D is close to 2 mm and λ is 550 nm

$$x_{crit} = 1.00 \times 2 \times 10^{-3}/1.22 \times 5.50 \times 10^{-7} \text{ m}$$
$$= 3.3 \times 10^3 \text{ m or } 3.3 \text{ km}$$

So the value of x_{crit} is not constant but changes by a factor of about 4 times as the lighting conditions alter.

NOTES

(i) Rayleigh's criterion is an empirical relation. It does not apply to everyone. Some people can resolve two diffraction images at distances greater than x_{crit} (calculated above) whereas others cannot.

(ii) The shift of the peak sensitivity of the 'normal' eye is called the Purkinje effect.

(iii) The Airy formula is very similar in form to that of the single rectangular slit, but the derivation of the irradiance distribution is much more complicated (it involves Bessel functions).

(iv) It is a rare person indeed whose eyesight is only limited by diffraction. Generally, one or more of the other types of aberrations, such as coma or astigmatism, will cause x_{crit} to be less than the calculated value.

(v) The resultant irradiance of the two irradiance distributions for incoherent light sources can be found by direct addition. With coherent sources the resultant amplitude must be determined first, taking into account the appropriate phase difference. Then the resultant irradiance can be found using the relation

$$I \propto (\text{resultant amplitude})^2$$

Example 4.7

Determine the resolving power of a prism having a refracting angle of 60.0° and a base width of 5.0 cm at a wavelength of 600 nm. It is used with a spectrometer adjusted for minimum deviation. The refractive index of the prism material obeys Cauchy's relation

$$n = 1.570 + 1.000 \times 10^4 \lambda^{-2}$$

where the coefficient of λ^{-2} is expressed in nm^2.

The source is assumed to consist of wavelengths λ and $\lambda + \delta\lambda$ only. A parallel beam from the collimator falls on the prism. After refraction by the prism the two wavelengths form diffraction images of the slit in the second focal plane of the telescope objective. The two images may be just resolved when the principal maximum of one occurs in the same position in the field of view as the first minimum of the other. In other words Rayleigh's Criterion is being applied.

To arrive at an expression for the resolving power, Fermat's Principle of Least Time is applied to each wavelength. In essence, this principle says that for both wavelengths the optical distance from the object, i.e. the slit, to the image is identical. If the incident beam fills the whole of the face of the prism then the optical paths can be determined. It is only necessary to assume that both wavelengths travel the same geometrical base distance. Of course, the refractive index n of the prism varies with wavelength.

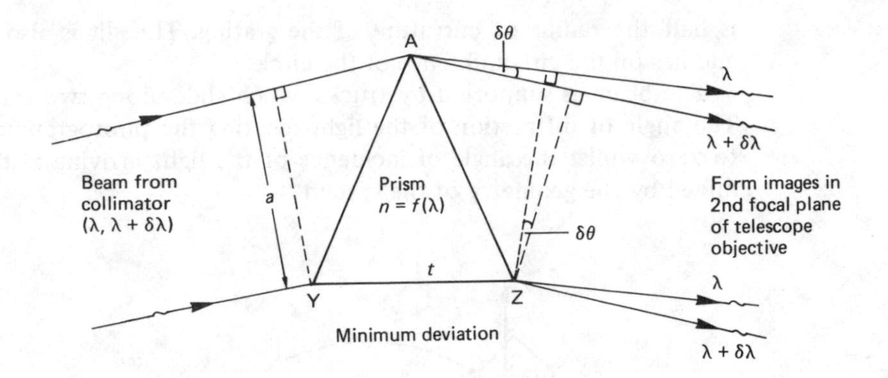

Data	*Given:*	constants in Cauchy relation: $A = 1.570, B = 1.000 \times 10^4$ nm^2
		base width t of prism = 5.0 cm
		refracting angle = 60.0°
	Unknown:	resolving power (RP)
		width of incident and emergent beams
Relevant equations	$\text{RP} = \lambda/\delta\lambda = t\,\delta n/\delta\lambda$	
	$\delta\theta = \lambda/a$	

Given that

$$n = A + B\lambda^{-2} \tag{1}$$

$$\delta n/\delta\lambda = -\,2B/\lambda^{-3} \tag{2}$$

and

$$\text{RP} = 2Bt/\lambda^{-3} \tag{3}$$

$$= 2 \times 1.000 \times 10^4 \times 10^{-18} \times 5.0 \times 10^{-2}/(6.00 \times 10^{-7})^3$$

$$= 4.6 \times 10^3$$

NOTES

(i) The resolving power of a prism is typically about 10^3, compared with a high quality plane transmission diffraction grating with a value of 10^5-10^6.

(ii) Care needs to be taken with the units. Generally questions contain mixed units which you must convert to a common unit. Here metres (m) are used.

Example 4.8

A concave diffraction grating of radius 6.00 m and having 6000 elements per cm is used in the Rowland mounting. A second order principal maximum is photographed at a distance of 1.80 m along the chord joining the illuminated slit and the photographic plate. Find the wavelength of the radiation being investigated.

There are basically two disadvantages of the plane diffraction grating: (a) a lens needs to be used to focus the various diffracted wave systems; (b) as the focal length of a lens depends on wavelength, not all the wavelengths can be focused simultaneously – even using an achromatic doublet does not eliminate chromatic aberration entirely.

The concave diffraction grating images the diffraction patterns on a photographic plate. In the Rowland mounting the grating and photographic plate are firmly fixed to a rigid beam when the distance between them is equal to the diameter of the 'Rowland Circle' (see below); the radius of the 'Rowland Circle' is half the radius of curvature of the grating. The slit is also fixed permanently and lies on the circumference of the circle.

The beam is supported by trucks which slide along two tracks at right angles. The angle of diffraction of the light meeting the photographic plate is kept close to zero whilst the angle of incidence of the light arriving at the grating is determined by the geometry of the system.

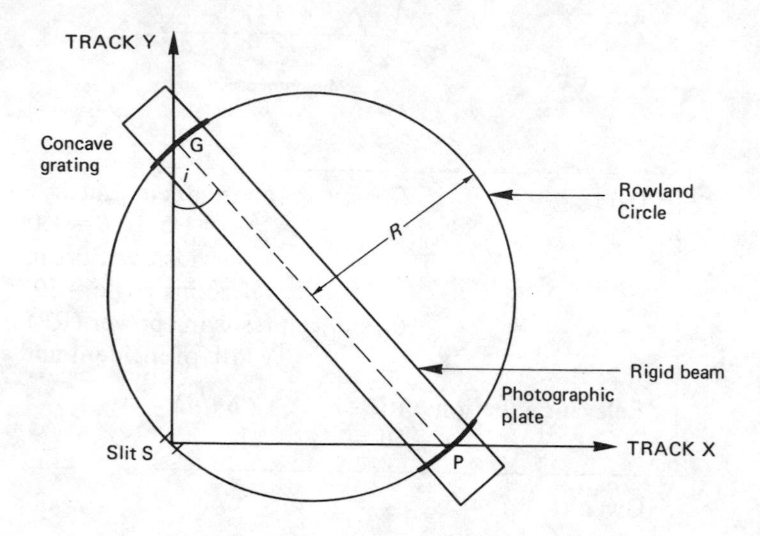

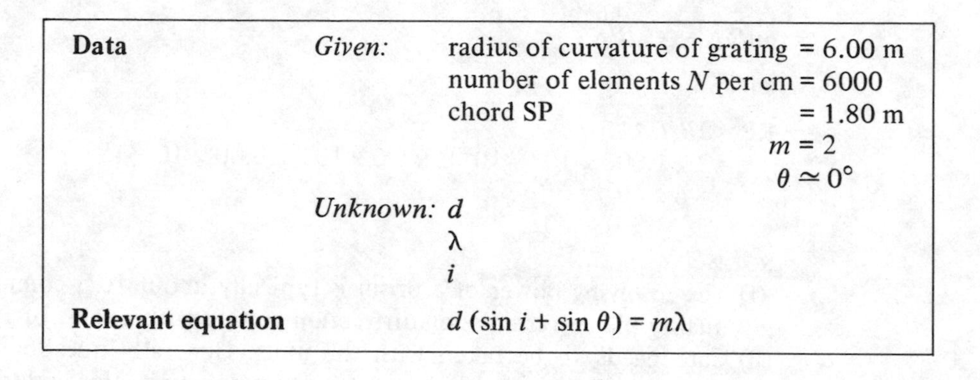

Data	*Given:*	radius of curvature of grating = 6.00 m
		number of elements N per cm = 6000
		chord SP = 1.80 m
		$m = 2$
		$\theta \simeq 0°$
	Unknown:	d
		λ
		i
Relevant equation		$d(\sin i + \sin\theta) = m\lambda$

As θ is arranged to be nearly zero, we can reduce the above relation to

$d \sin i = m\lambda$

Therefore

$$\lambda = (d/m)\sin i \qquad (1)$$

As

$d = 1/N$

and

$$\sin i = \text{PS/GP (from diagram)} \qquad (2)$$

then

$$\lambda = (1/Nm)(\text{PS/GP}) \qquad (3)$$
$$= 1.80/(6.00 \times 6000 \times 2)$$
$$= 2.50 \times 10^{-5} \text{ cm}$$

NOTES

(i) The calculated wavelength lies outside the visible spectrum in the near ultra-violet. This kind of grating has the advantage that wavelengths on either side of the visible can be studied without being absorbed by the grating material.

(ii) The experimental set-up is carefully designed to give an angle of diffraction close to zero. If there are a number of wavelengths in the source then a *normal* spectrum is produced on the photographic plate, i.e. a spectrum over which the linear dispersion is constant.

(iii) The maximum wavelength which can be studied within any given order is equal to d/m.

(iv) The major disadvantage of the Rowland mounting is its large physical size; temperature variations may occur between the various components of the mounting, and the grating interval may alter during the course of an experiment. The Eagle mounting is much smaller and does not suffer from the latter problem.

Example 4.9

A thin converging lens is used to obtain the diffraction pattern generated by a transmission cosinusoidal grating in its secondary focal plane. Calculate the linear separation between principal maxima if the focal length of the lens is 50.0 cm, the spatial period of the grating is 6.00×10^{-4} cm and the wavelength of the light used is 6.00×10^{-5} cm.

A cosinusoidal diffraction grating differs from a square-wave grating in that the transmitted light distribution across its plane varies sinusoidally. At its central point, where the optical axis intersects the grating, the light distribution takes a maximum value; this is the reason for the name 'cosinusoidal'. The amplitude transmittance t (i.e. the ratio of the transmitted amplitude to the incident amplitude) is defined by

$$t = \tfrac{1}{2}\,[1 + \cos(2\pi z/d)]$$

where the coordinate z is measured along the plane of the grating and d is the spatial period, i.e. the distance between neighbouring maxima in the light distribution. The spatial frequency q may also be used, defined as the number of pairs of maxima per unit length. Basic theory indicates that the Fraunhofer diffraction pattern consists of three principal maxima ($m = 0$ on the axis and $m = \pm 1$ off-axis) generated by three sets of diffracted wave systems.

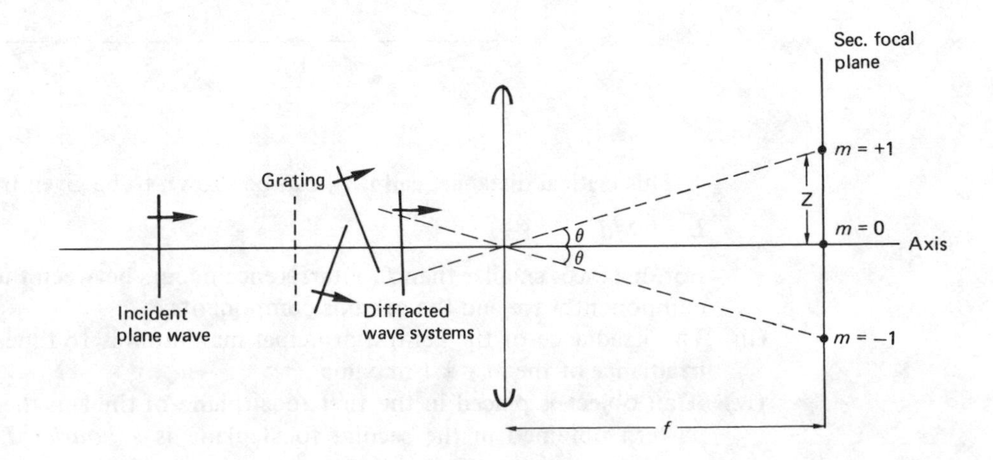

Data	Given:	$d = 6.00 \times 10^{-4}$ cm
		$\lambda = 6.00 \times 10^{-5}$ cm
		$f = 50.0$ cm
	Unknown:	θ
		z
Relevant equations		$d \sin\theta = \lambda$
		$\tan\theta = z/f$
		$\theta = \sin^{-1}(\lambda/d)$
		$\therefore z = f[\tan(\sin^{-1}(\lambda/d))]$

This is one example in which it is advisable to calculate the value of θ before determining the value of z. This is because θ may be a small angle (i.e. less than about 15°), when the above expression for z simplifies to

$$z = f\lambda/d \tag{1}$$

Now

$$\sin\theta = \lambda/d$$
$$= 6.00 \times 10^{-5}/6.00 \times 10^{-4}$$
$$= 1.00 \times 10^{-1}$$

and, therefore,

$$\theta = 5.7°$$

As this is, indeed, smaller than 15° we can use (1) and find

$$z = 50.0 \times 6.00 \times 10^{-5}/6.00 \times 10^{-4}$$
$$= 5.00 \text{ cm}$$

NOTES

(i) The incident light is assumed to be normal to the plane of the grating.

(ii) It is possible to observe the Fraunhofer diffraction pattern without using a lens. However, we must be sure to be beyond a critical distance from the grating at which the diffracted wave systems begin to separate (as shown in the diagram).

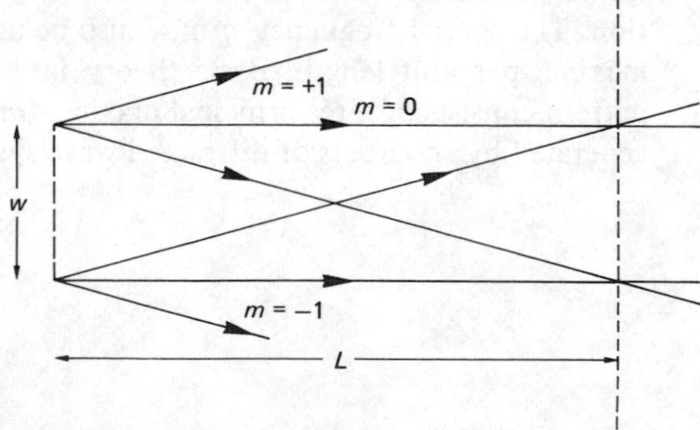

This critical distance, call it L, can be shown to be given by

$$L = w\lambda/d$$

For distances smaller than L, interference occurs between the zero frequency component wave and the off-axis components.

(iii) The irradiance of the central principal maximum is 16 times higher than the irradiance of the $m = \pm 1$ maxima.

(iv) If an object is placed in the first focal plane of the lens then the diffraction pattern obtained in the second focal plane is a *Fourier Transform* of the optical field across the object, i.e. the diffraction pattern contains information about the amplitude and phase at each point on the object. This part of optics is known as *Fourier Optics*.

(v) If the three plane-wave systems on the output side of the grating are extrapolated back to meet its back surface, then the original amplitude distribution will be obtained. Each wave system is characterised by a particular value of the spatial frequency q. The attenuated undiffracted wave moving along the axis has a spatial frequency equal to zero. It is sometimes called the DC component. For the other two waves, the fundamental frequency components have the spatial frequencies: $q_+ = +1/d$ and $q_- = -1/d$; the plus sign is generally used for the wave propagating above the axis.

The equation

$$d \sin \theta = \lambda$$

can be written

$$\sin \theta = q\lambda$$

So, the DC component has an equivalent spatial period of $1/0$ or ∞. In other words, this kind of wave can be generated by a uniformly 'fogged' photographic plate in which the grating element extends from $-\infty$ to $+\infty$. In general, every diffracting body can be thought of in terms of its spatial frequency components; the zero, fundamental and higher frequency components. Projecting the spatial frequency components back to the output surface of the diffracting body recreates the original amplitude distribution of the light across the object.

Example 4.10

X-rays of wavelength 1.542×10^{-10} m are incident on a crystal of rock salt at a glancing angle of 15°50′. Calculate the separation of the atomic layers assuming that a first order reflection occurs.

For diffraction to be experimentally observed with X-rays it is essential that the size d of the diffracting element is comparable with their wavelength. The only appropriate diffracting element for X-rays is the spacing between the atomic layers in a material.

In the ideal case of a crystalline material in which the atoms form lattice rows separated by a distance d, diffraction occurs when Bragg's law is obeyed, viz.

$$2d \sin \theta = m\lambda$$

The diffracted waves interfere to produce a set of spots on a photographic film or a high level signal which can be measured with a diffractometer.

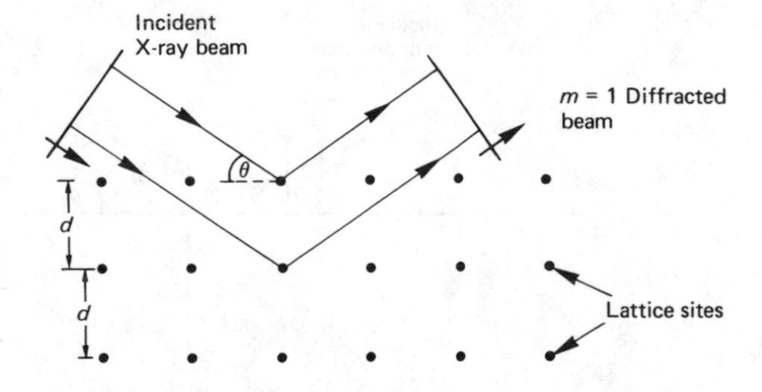

Data	*Given:*	$\theta = 15°50'$
		$m = 1$
		$\lambda = 1.542 \times 10^{-10}$ m
	Unknown:	d
Relevant equation		$2d \sin \theta = m\lambda$

$$d = 1 \times 1.542 \times 10^{-10}/2 \sin 15°50'$$
$$= 2.826 \times 10^{-10} \text{ m}$$

NOTES

(i) It is important to remember that in problems dealing with X-ray diffraction the angle of incidence is always measured with respect to the surface, as indicated in the diagram.

(ii) With polycrystalline materials, the diffraction pattern consists of concentric circles. These are produced by individual diffraction spots lying so close to one another that the end result is a continuous ring. In a powder diffraction camera the interference spots arise from the random arrangement of microcrystallites. With amorphous materials the diffraction pattern consists of a few diffused haloes.

Example 4.11

Monochromatic light is incident non-normally on a plane transmission diffraction grating having 1200 elements per mm. Two successive diffraction beams are observed to occur at 14.0° and 73.0° to the normal on the same side of the optical axis. Calculate the wavelength of the light.

Due to the light being incident non-normally there is an addition optical path difference that must now be taken into consideration on deriving the interference condition. If the angle of incidence is i then this additional OPD is $d \sin i$. The zero order principal maximum no longer lies on the optical axis of the system. As with the reflection-type grating, an algebraic convention is used to distinguish between those incident and diffracted beams which lie above or below the axis.

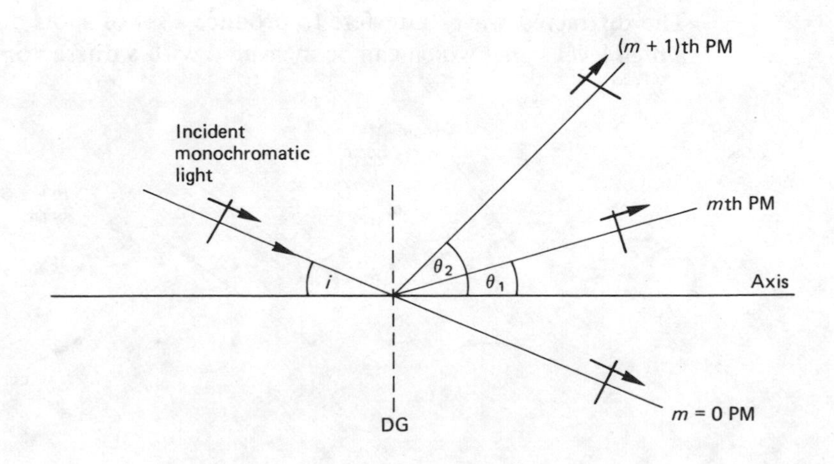

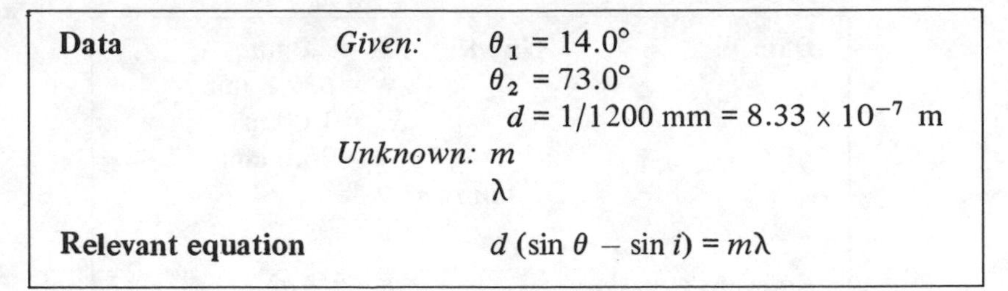

Data *Given:* $\theta_1 = 14.0°$
$\theta_2 = 73.0°$
$d = 1/1200 \text{ mm} = 8.33 \times 10^{-7}$ m
Unknown: m
λ

Relevant equation $d(\sin\theta - \sin i) = m\lambda$

The interference conditions for each principal maxima are:

$$d \sin\theta_1 - d \sin i = m\lambda \qquad (1)$$

and

$$d \sin\theta_2 - d \sin i = (m + 1)\lambda \qquad (2)$$

Subtracting (1) from (2) gives

$$\lambda = d(\sin\theta_2 - \sin\theta_1) \qquad (3)$$

$$= 8.33 \times 10^{-7} \times (\sin 73.0° - \sin 14.0°) \text{ m}$$
$$= 8.33 \times 0.7144 \times 10^{-7} \text{ m}$$
$$= 5.95 \times 10^{-7} \text{ m}$$

Example 4.12

A thin lens of focal length 25.0 cm has a circular aperture of 2.0 mm placed centrally and immediately in front of it. If an object is 1.00 m away from the lens determine the maximum spatial frequency component that will be reproduced within an image formed on the optical axis of the system using He-Ne laser light of wavelength 632.8 nm.

Using the spatial frequency technique introduced in Note (v) of Example 4.9, the object can be described in terms of the variation in light amplitude across its surface. This gives rise to a number of plane wave systems, characterised by different spatial frequencies q. The higher the value of q the larger the angle that the wave makes with the optical axis. Due to the finite size of the aperture, not all frequency components will be involved in image formation. This means that the image will not be a faithful reproduction of the object although it may still be possible to recognise it. As higher spatial frequencies correspond to smaller spatial periods, fine detail will be lost from the image.

Each frequency component will give rise to a disc of light in the image plane. The image will resemble the object only in those regions where the discs overlap. As the spatial frequency increases the discs move away from the axis. There will be a critical value of the spatial frequency q_c for which the region of overlap is a point. It is this value that we must calculate in this question.

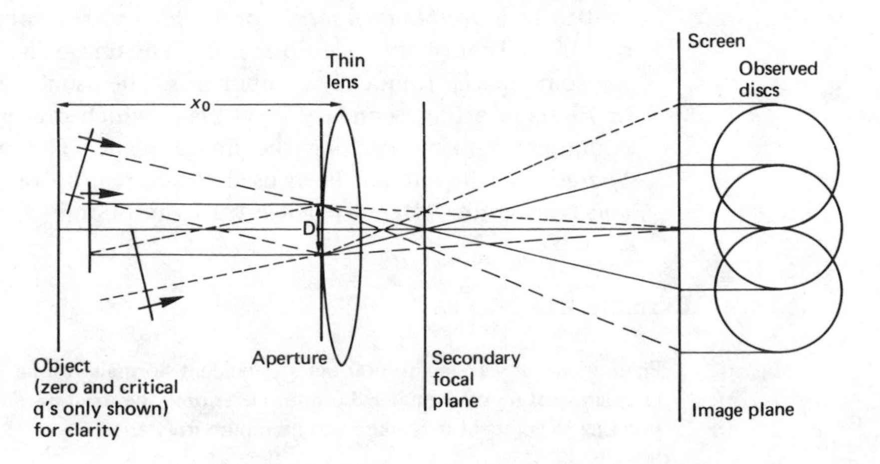

Data	*Given:*	$D = 2.0$ mm
		$\lambda = 632.8$ nm
		$x_o = 1.00$ m
		$f = 25.0$ cm
	Unknown:	θ
		q_c
Relevant equations		$\sin\theta = q\lambda$
		$D = 2x_o \tan\theta$

At the critical spatial frequency q_c the plane waves just pass through the aperture, therefore

$$\theta = \tan^{-1}(D/2x_o) \qquad (1)$$

and

$$q_c = \sin[\tan^{-1}(D/2x_o)]/\lambda \qquad (2)$$

(N.B. The largest value that q_c can have with the given lens occurs when x_o is equal to f, then

$$q_c = \sin[\tan^{-1}(D/2x_o)]/\lambda)$$

Now

$$D/2x_o = 2.0/2 \times 1.00 \times 10^3$$
$$= 1.0 \times 10^{-3}$$

As this is a small angle we can adopt the small angle approximation, then

$$q_c = D/2x_o\lambda$$
$$= 1.0 \times 10^{-3}/6.328 \times 10^{-7}\ \text{m}^{-1}$$
$$= 1.6 \times 10^3\ \text{m}^{-1}$$
$$= 1.6\ \text{mm}^{-1}$$

NOTES

(i) The largest *possible* value of q_c will occur for x_o equal to 25.0 cm, for if the object encroaches within the first focal plane the image is no longer real. q_c can be calculated to be 6.4 mm^{-1}.

(ii) The ratio of focal length to lens diameter is the *F*-number. It is possible to make high quality lenses for which F equals 1. Then, q_c is $1/2\lambda$.

(iii) The lens is a Fourier transform lens. It forms the Fraunhofer diffraction pattern in its secondary focal plane. Every point within this diffraction pattern may be regarded as a Huyghens' secondary source emitting wavelets which interfere in the image plane. In this context, then, image formation is said to be a *double diffraction* or *double transform* process. This is the basis of Abbe's Theory of Image Formation. The image can be altered by obstructing some spatial frequency components. The usual method is to insert masks or filters into the secondary focal plane which prevent particular frequency components from reaching the image plane. The technique is known as *Spatial Filtering.* It has been used to sharpen up photographs of the moon and remove unscattered particle tracks in bubble chamber photographs, for example.

Example 4.13

Plane waves of wavelength 600 nm are incident normally on an opaque screen possessing a circular aperture of diameter 3.0 mm. Determine the axial distances from the screen corresponding to points of maximum and minimum irradiance.

This is an example of Fresnel diffraction. With reference to a point P on the axis, the incident plane wavefront must be divided up into a number of half-period zones when it is in the plane of the aperture. Obviously, when P is a long way from the aperture it 'sees' less zones than when it is close to it. There will be a maximum irradiance when 1, 3, 5, . . ., $(2r - 1)$ zones fill the aperture and minimum irradiance when 2, 5, 6, . . ., $2r$ zones fill it (in both cases, r is an integer). It is assumed that the plane waves are produced by a source placed at infinity.

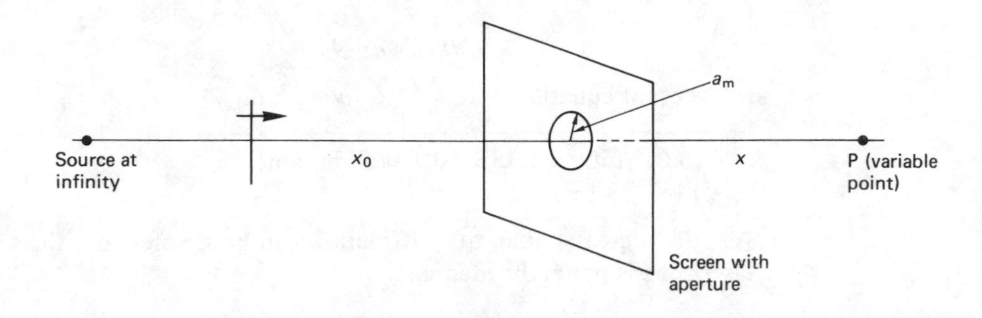

Data	*Given:*	$x_o = \infty$
		$a_m = 1.5$ mm
		$\lambda = 600$ nm
	Unknown:	x
		m
Relevant equation		$a_m^2 = m\lambda x x_o/(x + x_o)$

Re-write this expression as

$$a_m^2 = m\lambda x/(1 + x/x_o) \qquad (1)$$

Then because $x_o = \infty$, this reduces to

$$a_m^2 = m\lambda x$$

For $m = 1$: $x_1 = a_1^2/\lambda$
$= (1.5 \times 10^{-3})^2/6.00 \times 10^{-7}$
$= 3.75$ m

So when P is 3.75 m from the screen one half-period zone completely fills the aperture; the irradiance at P is a maximum.

When $m = 2$: $x_2 = a_2^2/2\lambda$
$= (1.5 \times 10^{-3})^2/2 \times 6.00 \times 10^{-7}$
$= 1.83$ m

Now two half-period zones fill the aperture and the irradiance at P is a minimum. And so on.

Example 4.14

An argon gas laser is constructed to operate at a wavelength of 488 nm. The resonant cavity is proposed to be 50.0 cm long and the reflecting mirrors are 6.0 mm diameter plane discs. Is this specification adequate to make diffraction losses negligible?

Gas lasers have open sides. One of the serious problems with them is a loss of energy through diffraction. Imagine a plane wave beginning at one mirror and travelling down the tube. A simple application of Huyghens' principle should enable you to see that some of the light must 'leak' out of the sides of the tube.

The Fresnel Number N is the important parameter here. If N is greater than about 50 then diffraction effects can be ignored. What this means is that an observer situated at one mirror can 'see' 50 or so half-period zones, and that to all intents and purposes radiation propagates down the cavity according to the laws of geometrical optics.

Data	*Given:*	$\lambda = 488$ nm
		$a = 6.0$ mm
		$L = 50.0$ cm
	Unknown:	N
Relevant equation		$N = a^2/L\lambda$

$$N = (6.0)^2/(500 \times 4.88 \times 10^{-4}) \text{ (all in mm)}$$
$$= 148$$

As 148 is greater than 50, diffraction can be ignored and the specification of the laser tube is perfectly adequate.

Example 4.15

An expanded beam of He-Ne laser light (wavelength 632.8 nm) is incident normally on a zone plate, 2.0 cm in diameter. It is determined experimentally that a bright diffraction spot is produced at an axial distance of 1.00 m from the zone plate. Determine the distance where the laser light will be focused if it *now* diverges from an axial point, 2.00 m in front of the zone plate.

A zone plate is a filter made by drawing a series of concentric circles on a transparent material with radii proportional to $m^{\frac{1}{2}}$ where m is an integer) and blackening either the odd regions or the even ones. Then the light transmitted by the clear regions will arrive at a point on the axis in phase. The irradiance will therefore be a maximum at this point. In this sense the zone plate acts like a lens; plane waves will be focused to a point and its distance from the zone plate is equivalent to the focal length.

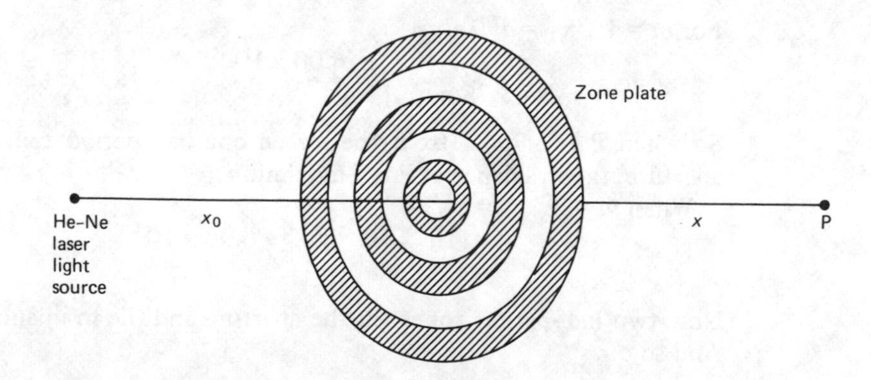

Data	*Given:*	(a) $a_m = 1.00$ mm	(b) $a_m = 1.00$ mm
		$x_o = \infty$	$x_o = 2.00$ m
		$x = 1.00$ m	$\lambda = 6.328 \times 10^{-7}$ m
		$\lambda = 6.328 \times 10^{-7}$ m	
	Unknown:	(a) m	(b) x
			m
Relevant equation		$a_m^2 = m\lambda x x_o/(x_o + x)$	

Re-write this expression as

$$\frac{1}{x_o} + \frac{1}{x} = \frac{m\lambda}{a_m^2} \qquad (1)$$

Now there is a strong resemblance to the thin lens equation:

$$\frac{1}{x_o} + \frac{1}{x} = \frac{1}{f} \qquad (2)$$

With $x_o = \infty$,

$$x = a_m^2/m\lambda \equiv f \qquad (3)$$

Thus x can be determined for other object positions.

f must be 1.00 m, therefore

for $x_o = 2.00$ m, we have

$$\frac{1}{x} + \frac{1}{2.00} = \frac{1}{1.00}$$

giving

$x = 2.00$ m

NOTES

(i) $a_m^2/m\lambda$ is the same as a_1^2/λ, putting m equal to 1.

(ii) Other images will be obtained at difference points on the axis. These correspond to $m = 3, 5, \ldots$, etc. Take m equal to 3, say – each zone is now divided into three sub-zones. The light from two of these sub-zones are out of phase by π leaving the effect of the third. A similar explanation can be used with m equal to 5, 7, etc.

Example 4.16

An aperture in an opaque screen has the shape shown in the following diagram. If plane waves of wavelength 500 nm are incident normally on the aperture, determine the amplitude of the light at an axial point 2.00 m from the screen in terms of the unobstructed amplitude.

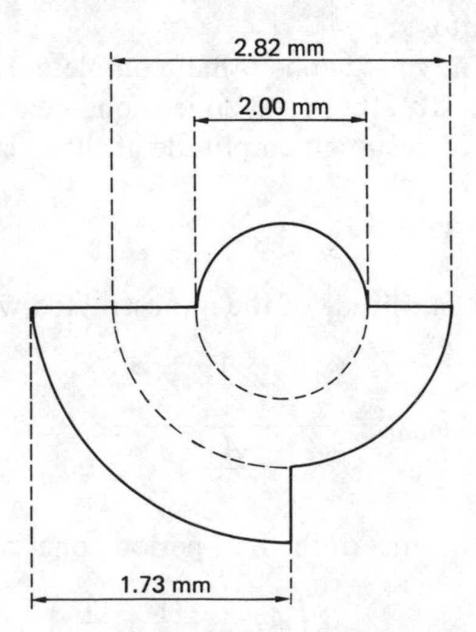

When the plane wavefront reaches the aperture every point on the unobstructed portion acts as a Huyghens' secondary source. The number of such sources is proportional to the zonal area. From the given axial point it is necessary to determine how many half-period zones are present within the aperture, then the resultant amplitude at this point can be found using

$$A = A_1 - A_2 + A_3 - A_4 + \ldots$$

modified to take account of the degree of obstruction of the zones. With plane wavefronts the amplitude of the unobstructed wave is $A_1/2$.

Data	*Given:*	radius of inner circle = 1.00 mm
		radius of semi-circle = 1.41 mm
		radius of quadrant = 1.73 mm
	Unknown:	amplitude A at axial point
Relevant equations		$m = a_m^2 (x + x_o)/xx_o\lambda$
		$A = A_1 - A_2 + A_3 - \ldots$
		$(A)_{unob} = A_1/2$

With plane waves, $x_o = \infty$, therefore

$$m = a_m^2/x\lambda$$

Therefore the number of zones occupying the various regions can be calculated.

For the inner circle:

$$m = (1.00 \times 10^{-3})^2/2.00 \times 5.00 \times 10^{-7}$$
$$= 1$$

For the semi-circle:

$$m = (1.41 \times 10^{-3})^2/2.00 \times 5.00 \times 10^{-7}$$
$$= 2$$

For the outer circle

$$m = (1.73 \times 10^{-3})^2/2.00 \times 5.00 \times 10^{-7}$$
$$= 3$$

The $m = 1$ zone is complete. Hence its contribution to the amplitude at the axial point is A_1.

The $m = 2$ zone is half complete. Hence it contributes an amplitude $A_2/2$.

Lastly, the $m = 3$ zone is one quarter complete. Its contribution is $A_3/4$.

The resultant amplitude at the axial point is thus

$$A = A_1 - \frac{A_2}{2} + \frac{A_3}{4}$$

The amplitude of the unobstructed wave is $A_1/2$. Therefore

$$A/(A)_{unob} = \frac{A_1 - \frac{A_2}{2} + \frac{A_3}{4}}{\frac{A_1}{2}}$$

As the area of the half-period zones are approximately equal, we can write

$$A/(A)_{unob} = \frac{2\left(A_1 - \frac{A_1}{2} + \frac{A_1}{4}\right)}{A_1}$$
$$= 3/2$$

Example 4.17

A single slit is 2.00 m from a point source of light and 4.00 m from a screen. If light of wavelength 600 nm is used, determine the value of the slit-width which produces an optical path difference of $\lambda/10$ between the peripheral and central waves. Find the irradiance at the central point on the screen relative to the irradiance of the unobstructed wavefront.

Given the optical path difference between the waves, the phase difference δ can be found, and, hence, ν. The slit width may now be calculated from the known values of x_o, x and λ. The Cornu Spiral of Fig. 4.5 can be used to obtain the resultant amplitude at the screen.

Data	*Given:*	$x_o = 2.00$ m
		$x = 4.00$ m
		$\lambda = 600$ nm
		OPD $= \lambda/10$
	Unknown:	slit width ($= 2a_m$)
Relevant equations		$\delta = (\pi/2)\,\nu^2 = (2\pi/\lambda)$OPD
		$\nu = a_m\ [2\,(x_o + x)/x_o x\lambda]^{\frac{1}{2}}$

Now

$$\delta = (2\pi/\lambda).\ \text{OPD} \tag{1}$$

Therefore

$$\nu = (4 \times \text{OPD}/\lambda)^{\frac{1}{2}} \tag{2}$$

and

$$a_m = [2x_o x\ \text{OPD}/(x_o + x)]^{\frac{1}{2}} \tag{3}$$

$$= 2 \times 2.00 \times 4.00 \times 6.00 \times 10^{-7}/10 \times 6.00\ \text{m}$$

$$= 4.00 \times 10^{-4}\ \text{m}$$

$$= 0.40\ \text{mm}$$

The slit width is 0.80 mm.

As ν can be calculated to be $\pm$ 0.63 units, measure this distance along the arc of each branch of the Cornu Spiral beginning at the origin. The distance between the end points gives the resultant amplitude. This is 1.27 units.

As the irradiance of the unobstructed wavefront is 2 units, the relative irradiance is $(1.27)^2/2$, equal to 0.82.

NOTE

If plane waves are incident on the slit, then ν is given by

$$\nu^2 = a_m^2\ (2/x\lambda)$$

Example 4.18

Using the single slit of the previous example, determine the irradiance at points 0.10 mm above and below the central axis on the screen – but using a source of plane waves.

When the observing point is below the axis it is necessary to translate the slit downwards until the point lies on the axis. The new coordinates of the slit will enable two values of ν to be calculated. The Cornu Spiral of Fig. 4.5 may then be used to determine the resultant amplitude. The same basic procedure applies to the observing point above the axis, except that the slit is translated upwards.

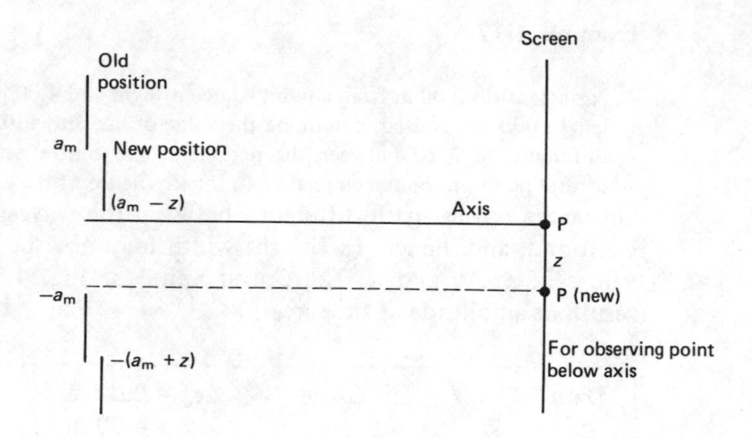

Data	*Given:*	as in Example 4.17
	Unknown:	irradiance 0.1 mm above and below the axis
Relevant equation		$\nu^2 = 2a_m^2/x\lambda$

Below the axis
New coordinates of the slit are $(a_m - z)$ and $-(a_m + z)$. Therefore

$$\nu_1 = (a_m - z).(2/x\lambda)^{\frac{1}{2}}$$

and (1)

$$\nu_2 = -(a_m + z).(2/x\lambda)^{\frac{1}{2}}$$

Hence

$$\nu_1 = 0.30 \times (0.83)^{\frac{1}{2}} = 0.27$$

and

$$\nu_2 = -0.50 \times (0.83)^{\frac{1}{2}} = -0.46$$

Locate the end points of these arcs on the Cornu Spiral. The length of the connecting line is 0.73 units. Therefore the relative irradiance is 0.27 units.

Above the axis
The new coordinates of the slit are $(a_m + z)$ and $(-a_m + z)$. Therefore

$$\nu_1 = (a_m + z).(2/x\lambda)^{\frac{1}{2}}$$

and (2)

$$\nu_2 = (-a_m + z).(2/x\lambda)^{\frac{1}{2}}$$

Hence

$$\nu_1 = 0.50(0.83)^{\frac{1}{2}} = 0.46$$

and

$$\nu_2 = -0.3(0.83)^{\frac{1}{2}} = -0.27$$

Thus the length of the connecting line on the Cornu Spiral is 0.73 units, as in case (a), and the normalised irradiance has the same value also.

NOTE
There is nothing startling about these results. From symmetry considerations, it is what we should have expected.

Example 4.19

A long wire, 0.50 mm in diameter, is placed 2.00 m from a point source of light (wavelength 500 nm). Determine the irradiance at an axial point 3.00 m from the wire.

The wire acts as an obstacle preventing a number of the central half-period strips from contributing to the resultant amplitude at the axial point. As the wire is long we need only concern ourselves with the effect of the diameter. Suppose that the wire lies along the Y-axis then values of ν corresponding to $z = +R$ and $z = -R$ may be found, as in the previous examples. Starting at the origin of the Cornu Spiral, travel along the arc of the spiral in both directions. Then the region between the termination points need not be considered. Draw phasors from Z and Z' to these points. Now draw a reduced vector diagram and obtain the resultant amplitude.

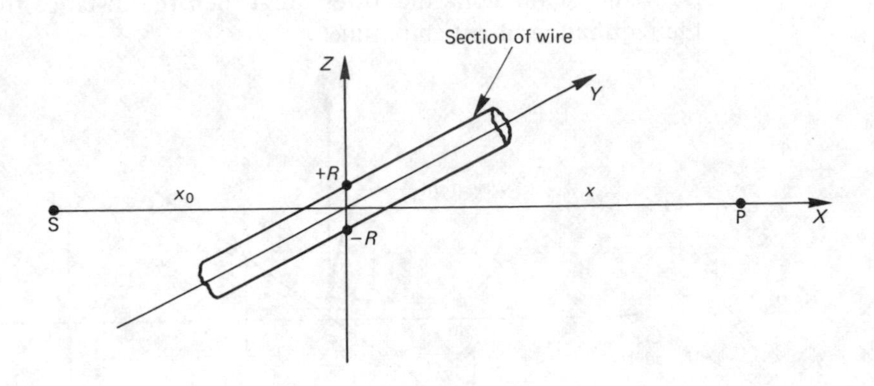

Data	*Given:*	$x_o = 2.00$ m $x = 3.00$ m $d = 0.50$ mm $\lambda = 500$ nm
	Unknown:	number of zones obscured resultant amplitude (and irradiance) at P
Relevant equation		$\nu = \pm a_m \; [2\,(x_o + x)/x_o x \lambda]^{\frac{1}{2}}$

$$\nu = \pm 2.50 \times 10^{-4} \; (2 \times 5.00/2.00 \times 3.00 \times 5.00 \times 10^{-7})^{\frac{1}{2}}$$
$$= \pm 0.45$$

Using the Cornu Spiral of Fig. 4.5 we find that the length of the two vectors is about 0.45 units and that the vectors lie along the same direction. The reduced vector diagram is simply a straight line of length 0.90 units. Therefore the resultant amplitude at P is 0.90 units and the irradiance is $(0.90)^2/2$ relative to the unobstructed wavefront, i.e. 0.41.

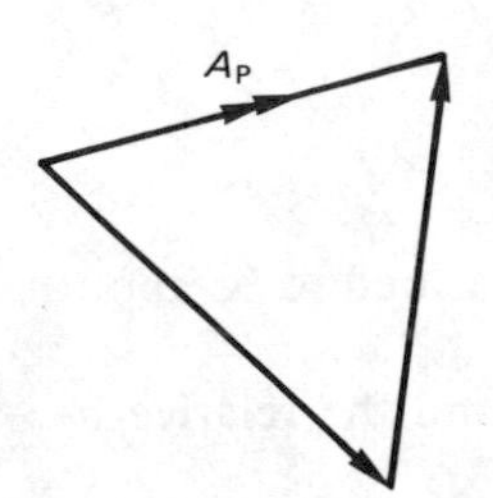

NOTES

(i) Due to the symmetry of the arrangement it should not be surprising that the vectors are parallel.

(ii) If the wire is displaced relative to the Y-axis then the length of the vectors would be different and they would point in different directions. Thus the reduced vector diagram might look like that on the left.

Example 4.20

A screen is 4.00 m from a point source of light of wavelength 600 nm. Midway between them is a knife edge. Determine the irradiance at (a) the point where the axis meets the screen and (b) 1.00 mm above and below this point.

If the point where the axis meets the screen is called P_o, then as 'seen' from P_o only one half of the incident wavefront would contribute to the irradiance. So, immediately, we could write down the relative irradiance as (length of phasor $Z'O)^2/2$, i.e. $\frac{1}{4}$.

For the points 1.00 mm above and below P_o, we need to determine the corresponding distance measured along the wavefront; this is a_m. Then convert it to its ν value. For the point above P_o measure a distance ν along the upper part of the Cornu Spiral. The distance from Z' to this point is the resultant amplitude (in the same way as before). For the point P below the axis, measure a distance ν from the origin along the lower spiral then the distance from Z' to this point is the required resultant amplitude.

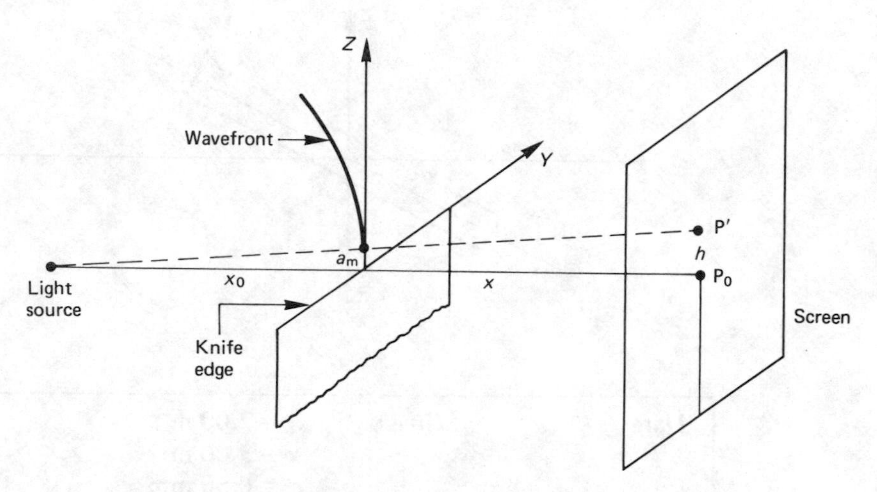

Data	*Given:*	$x_o = 2.00$ m
		$x = 2.00$ m
		$\lambda = 600$ nm
		$h = \pm 1.00$ mm
	Unknown:	a_m
		ν
		$A_{P'}$ (above and below axis)
Relevant equations		$a_m = x_o h/(x_o + x)$ (similar triangles)
		$a_m = \nu\,(x_o x\lambda/2\,(x_o + x))^{\frac{1}{2}}$

Now

$$a_m = 2.00 \times 1.00 \times 10^{-3}/4.00$$
$$= 0.50 \times 10^{-3} \text{ m}$$

and

$$\nu = 5.0 \times 10^{-4}\ (2 \times 4.00/2.00 \times 2.00 \times 6 \times 10^{-7})^{\frac{1}{2}}$$
$$= 0.9$$

For P′ above the axis, the resultant amplitude may now be measured to be about 1.5 units and the relative irradiance is $(1.5)^2/2$, i.e. 1.1.

For P′ below the axis the resultant amplitude is 0.3 units and the relative irradiance is $(0.3)^2/2$, i.e. 0.045.

NOTES

(i) As P′ penetrates further into the geometric shadow the resultant amplitude falls off continuously to zero.

(ii) As P′ moves further above the axis the resultant amplitude oscillates about a mean value of $2^{\frac{1}{2}}$, finally attaining this value when it is far enough away from the axis, for then it 'sees' the unobstructed wavefront. This result is shown in the diagram.

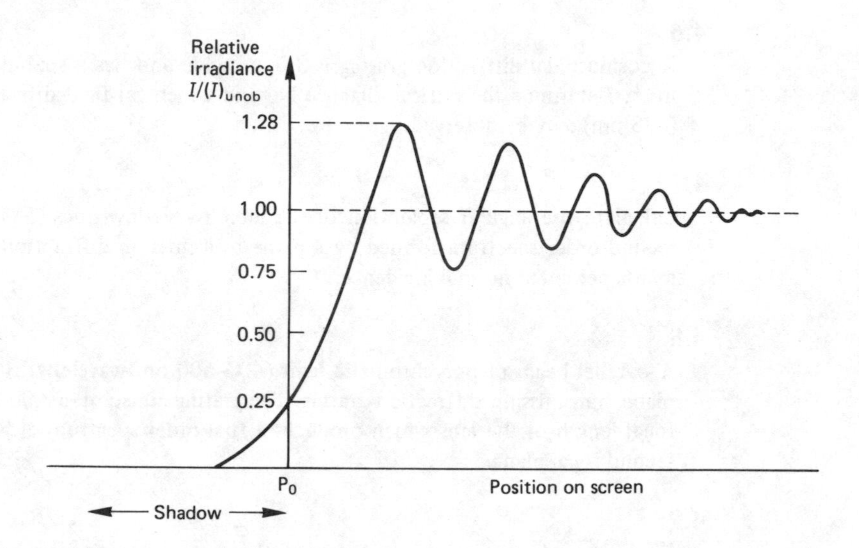

4.6 Questions

4.1

The visual acuity of the human eye in daylight is set by the density of cones in the foveal region, viz. 1.4×10^5 mm^{-2}. Show that if the distance between the eye lens and the retina is 17.0 mm the smallest angle subtended by an object at the eye is about 0.5 minutes of arc.

4.2

In the single slit diffraction pattern for white light the 4th minimum for green light (500 nm) coincides with the 3rd minimum for an unknown wavelength. Calculate its value.

4.3

Two parallel vertical slits are each 0.5 mm wide. Their near sides are 2.0 mm apart. Sketch a vector diagram for the case in which the phase difference between identically situated points in the two slits is $2\pi/5$. Obtain the value of half the phase difference across one of the slits.

4.4

A plane transmission diffraction grating of width 1.5 inch has 7500 elements inch^{-1}. Yellow sodium light of wavelengths 589.0 nm and 589.6 nm is incident on the grating. Find:

(a) The maximum order of interference produced.
(b) The smallest wavelength that may be detected in the 2nd order spectrum.
(c) The angular dispersion in the 3rd order spectrum.
(d) What additional information do you require in order to determine whether missing orders occur?

4.5

A concave diffraction grating of radius R has N elements per cm. Light from a slit situated on the Rowland circle is incident on the grating at an angle of incidence i. The mth order diffracted light of wavelength λ comes to a focus on the Rowland circle at a distance L along its arc from the pole of the grating. If i' is the angle of diffraction show that

$$L = R\left(\frac{\pi}{2} - i'\right)$$

and

$$\mathrm{d}L/\mathrm{d}\lambda = NmR/\cos i'$$

4.6

A cosinusoidal diffraction grating is 3.0 cm wide and has a spatial frequency of 1.66×10^4 m^{-1}. Determine the critical distance beyond which far-field diffraction of He-Ne radiation (633 nm) may be observed.

4.7

Calculate the angular separation of the mercury yellow lines (577.0 and 579.1 nm) in the second order spectrum formed by a plane transmission diffraction grating having 1000 elements per cm at normal incidence.

4.8

A parallel beam of polychromatic light (400-600 nm wavelength) is incident normally on a plane transmission diffraction grating. The grating constant is 2.00×10^{-3} cm. Calculate the focal length of the lens which produces a first order spectrum of linear width 3.00 cm in its second focal plane.

4.9

Find the radius of the central disc of the Airy diffraction image of a star formed by the objective lens of a camera of 2.5 cm diameter and 7.5 cm focal length.

4.10

Waves from a distant source have a plane wavefront. They are incident on an opaque screen in which there is a circular aperture. As seen from an axial point P the aperture consists of one half of the first Fresnel half-period zone. Determine the irradiance at P in terms of the irradiance of the unobstructed wavefront.

4.11

An aperture in an opaque screen has the shape shown in the diagram. Plane waves of amplitude 2 units and wavelength 500 nm are incident on the aperture. Determine the amplitude at an axial point, 2.00 m from the screen. What is the phase relation between the optical field at this point with and without the screen?

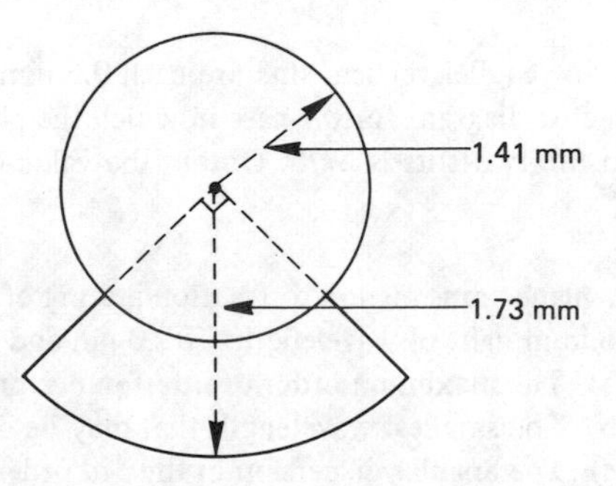

4.12

Huyghens' principle says that:

(a) for a given wavelength every point on a wavefront passing through an aperture acts as a secondary source;

(b) there are no wavelets in the 'backward' direction;

(c) the wavefronts are spherical in optically homogeneous materials;
(d) every point on a wavefront acts as a source of secondary wavelets and the envelope of all these wavelets at some later time gives the new position of the wavefront;
(e) none of these but . . .

4.13

Diffraction occurs:
(a) when waves pass through a slit of width much smaller than the wavelength;
(b) as a consequence of Huyghens' principle;
(c) with light waves only;
(d) because the Huyghens–Fresnel theory predicts it;
(e) none of these but . . .

4.14

When light of wavelength λ is incident on a long narrow slit of width a, the main maximum in the diffraction pattern has an angular size of:
(a) $\lambda/2a$;
(b) $2\lambda/a$;
(c) $2\pi a/\lambda$;
(d) $a/2\lambda$;
(e) none of these but . . .

4.15

Missing orders in a diffraction pattern occur because:
(a) a single slit is too wide;
(b) the mark-space ratio is zero or an integer;
(c) the grating element is not the same across a diffraction grating;
(d) the mark-space ratio is integral;
(e) none of these but . . .

4.16

Rayleigh's criterion:
(a) says that the Airy discs must overlap in the focal plane of an observing lens;
(b) has a theoretical basis and applies to everyone;
(c) says that two circular diffraction images formed by a lens of diameter D will be just resolved when the angle between their centres is greater than $1.22\lambda/D$ and the normalised irradiance between them falls to 0.81;
(d) can only be used on an empirical basis but applies to everyone;
(e) none of these but . . .

4.17

If an object is placed in the first focal plane of a lens:
(a) the diffraction pattern is its image;
(b) the image is the Fourier transform of the optical field across the object;
(c) the diffraction pattern will be observed in the second focal plane and will always be an infinitesimally sized spot;
(d) the diffraction pattern obtained in the secondary focal plane is a Fourier transform of the amplitude/phase relation across the object;
(e) none of these but . . .

4.18

An object may be regarded as a set of spatial frequency components:
(a) which must all be captured by a lens in order for the image to faithfully replicate the object;
(b) which describes the variation in light intensity across it;
(c) which produces a set of circular discs in the focal plane of an observing lens;
(d) but the frequency component given by $q = \sin\theta/\lambda$ is the important one;
(e) none of these but . . .

4.19

A zone plate:

(a) is essential for Fresnel diffraction to occur;

(b) is a filter made by constructing a set of black and white annuli with diameters proportional to the order of interference m;

(c) is equivalent to a thin lens with a variable focal length;

(d) allows either the even or the odd Fresnel half-period zones to be brought to a focus at various points on an axis drawn perpendicular to its plane;

(e) none of these but . . .

4.20

When light is diffracted by straight edges and slits the emerging wavefront:

(a) must always have a resultant amplitude given by the sum or difference of half the amplitudes of the first and last half-period zones;

(b) is cylindrical and must be divided into an infinite number of half-period zones;

(c) must be divided into half-period strips and the resultant amplitude at some external point can be obtained using a vector diagram called the Cornu Spiral;

(d) is called the Cornu Spiral from which the resultant amplitude can be calculated;

(e) none of these but . . .

4.21

A transmission-type diffraction grating cannot be used with ultra-violet radiation because:

(a) diffraction only occurs in the visible;

(b) it becomes heated and the grating element alters in size;

(c) it is difficult to detect the diffraction pattern;

(d) reflection gratings are always used for this purpose;

(e) none of these but . . .

4.7 Answers to Questions

4.1 0.5 minutes of arc

Calculate the separation between foveal cones and then the angle subtended at the centre of the eye lens. If two objects subtend an angle less than 0.5 min at the eye they will not be resolved.

4.2 667 nm

This is a simple application of: $a \sin \theta = m\lambda$ for minima.

4.3

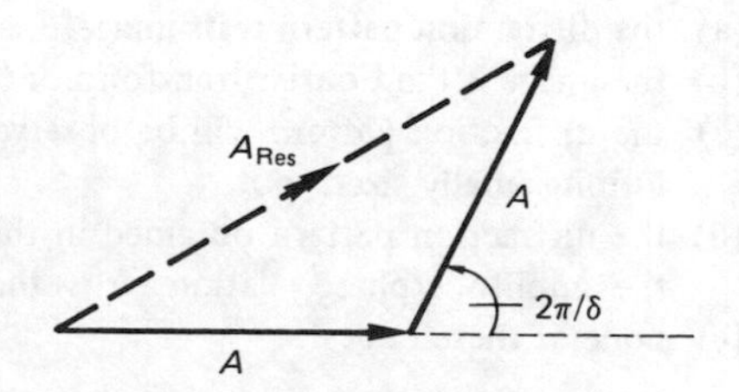

The phase difference 2β across one slit is $2\pi a/5\ (a + b) = 2\pi/25$.

4.4 (a) 5

Put $\sin \theta = 1$ in the interference condition.

(b) 2.6×10^{-11} m

Resolving power is defined as $\lambda/\Delta\lambda = mN$

(c) 1.04×10^6 m^{-1}

The angular dispersion is defined as the change in the angle of diffraction brought about by a small change in wavelength, i.e. $\Delta\theta/\Delta\lambda$. It can be shown to equal $m/d \cos\theta$. θ is first found from the usual diffraction relation;

$d \sin\theta = m\lambda$

(d) The width of each transparent region.

4.5 The first part utilises the geometry of the circle.
The second part starts with the relation:

$d(\sin i + \sin i') = m\lambda$

Differentiate with respect to λ and use the fact that

$dL/di' = (dL/d\lambda) \times (d\lambda/di')$

The negative sign that appears indicates that the focus position increases as the wavelength investigated decreases.

4.6 3.2 m
Use the relation: $L = w\lambda/d$, given in Example 4.9.

4.7 4.2×10^{-4} radians
Do not assume a small-angle approximation. Calculate the angle of diffraction for each wavelength using the standard grating formula. If the angles are less than about 15° they can be left in radians.

4.8 150 cm
Determine the angles of diffraction, as in Question 4.7. Then use the technique adopted in Example 4.5 to determine the focal length.

4.9 1.8×10^{-4} cm
The angle of diffraction for the central disc of the Airy diffraction image is given by $1.22\lambda/D$, where D is the diameter of the objective. This can be equated with r/f, where r is the required radius and f the objective's focal length. The only remaining problem is the selection of a value for the wavelength. In obtaining the above answer, a wavelength of 500 nm has been used, i.e. the wavelength at which the eye has its peak sensitivity under twilight conditions.

4.10 $I_P = I_{un}$
One half of the first half-period zone is involved. So as the number of secondary sources is proportional to area, the amplitude at P is $A_1/2$, giving an irradiance $= kA_1^2/4$. The amplitude of the unobstructed wavefront is $A_1/2$, giving the same irradiance.

4.11 $A_1/4$; in phase
Use the relation

$a_m = [m\lambda x/(1 + x/x_o)]^{\frac{1}{2}}$

For plane waves $x_o = \infty$, when the relation reduces to

$a_m = (m\lambda x)^{\frac{1}{2}}$

Determine how many half-period zones are contained within the given shape. When $m = 2$, $a_2 = 1.41$ mm – the radius of the central area. Therefore it contains two half-period zones. When $m = 3$, $a_3 = 1.73$ mm – the radius of the quadrant. Thus

$$A_p = A_1 - A_2 + A_3/4$$

As the areas are about equal, A_p is $A_1/4$.

The waves from the periphery of the second zone are in anti-phase with those from the first zone, and the waves from the third zone are in phase with the first. Hence the phase of the optical field at P is exactly the same as with an unobstructed wavefront.

4.12 (d)

4.13 (a)

4.14 (b)

4.15 (d)

4.16 (c)

4.17 (d)

4.18 (a)

4.19 (d)

4.20 (c)

4.21 (e)
If the transmission grating is made of glass or some plastics then UV will be absorbed.

5 Polarisation

5.1 Representation of Unpolarised and Plane-polarised Light

In Section 2.1 it was demonstrated that if the free end of a rope is 'flicked' in a completely random manner then an unpolarised wave motion will be produced. On the other hand if the free end is 'flicked' in the same direction a plane-polarised wave will be formed. Figure 5.1 is one way of representing these two cases. You, the observer, are looking towards the source of the waves. The numbers in Fig. 5.1(a) refer to the order in which the individual plane-polarised waves are generated. It must be emphasised that the process is a *completely arbitrary* one and that over a long period of time all directions are equally likely. In Fig. 5.1(b), however, the waves *always* have the same plane of polarisation.

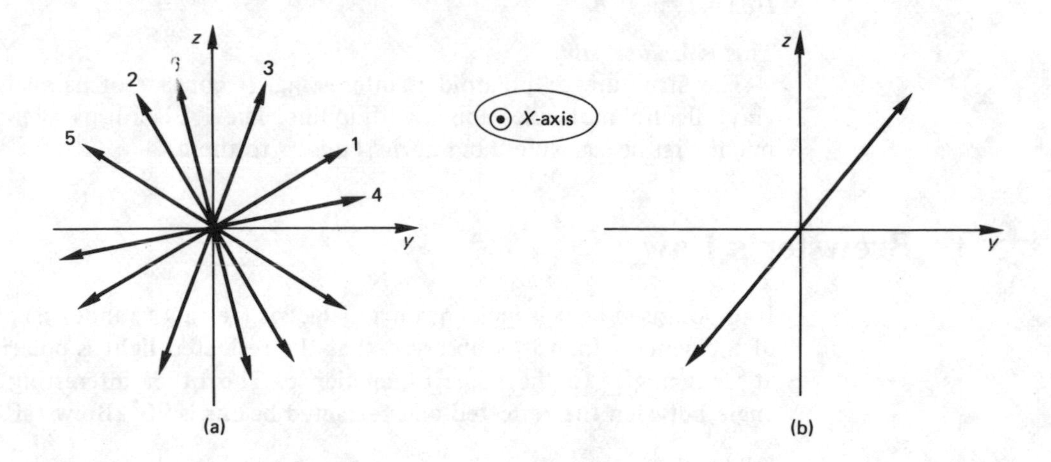

Figure 5.1

For convenience we shall refer to the vibrating quantity in a wave as the *optical vector*. It is actually the electric (E)-vector of the electromagnetic wave.

5.2 Transmission through Polaroid—Dichroism

If the optical vector of a light wave is parallel to a certain preferred direction in the polaroid, called the transmission axis (TA), then the light will be transmitted without appreciable absorption. If the optical vector is at right angles to the TA then no light will be transmitted. In Fig. 5.2 only the component OB is transmitted.

Let I_o be the irradiance of unpolarised light incident on polaroid. Now resolve each optical vector in Fig. 5.1(a) into two components, parallel and perpendicular to the TA. As there will be just as many vectors along each of these directions this means that the irradiance of each component is $I_o/2$. This is the maximum irradiance that the polaroid can transmit. Even if the polaroid is rotated there will be

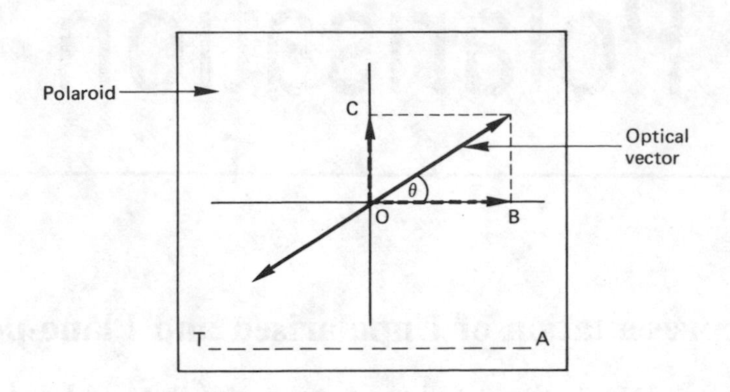

Figure 5.2

no change in the transmitted irradiance. The transmitted light is plane or linearly polarised. It is sometimes said to be in a *P-state* of polarisation.

If plane-polarised light of irradiance I_o is incident on polaroid then it will be transmitted fully if its optical vector is parallel to the TA but if there is an angle θ between them then the transmitted irradiance $I(\theta)$ is given by

$$I(\theta) = I_o \cos^2 \theta \tag{5.1}$$

This is *Malus's law*.

The structure of polaroid is interesting. It consists of parallel chains of poly-vinyl alcohol molecules doped with iodine. The TA is not parallel to the chains, as might first be suspected, but at *right angles* to them.

5.3 Brewster's Law

If unpolarised light is incident on a dielectric (refractive index n) at a certain angle of incidence $\bar{i}$ then it is observed that the reflected light is polarised in the plane at right angles to the plane of incidence. The other interesting fact is that the angle between the reflected and refracted beams is 90°. Brewster's law states that

$$\tan \bar{i} = n \tag{5.2}$$

if medium 1 is air or vacuum. As $n = f(\lambda)$, $\bar{i} = f(\lambda)$.

5.4 Superposition of Orthogonal Wave Motions

The worked examples given in Section 3.5 were all concerned with interference between waves polarised in the *same* plane. Suppose, now, that waves of amplitudes A_Y and A_Z are polarised along the Y- and Z-axes of a right-handed Cartesian coordinate system and propagating along the X-axis, as in Fig. 5.3.

We will represent the waves by

$$\left.\begin{aligned} E_Y &= A_Y \cos(kx - \omega t + \phi_1) \\ &\text{and} \\ E_Z &= A_Z \cos(kx - \omega t + \phi_2) \end{aligned}\right\} \tag{5.3}$$

where k is the wave-number $2\pi/\lambda$, ω is the angular frequency and the phase difference $\delta = (\phi_2 - \phi_1)$. Equation (5.3) tells us that the Z-wave *lags* the Y-wave by this phase difference δ.

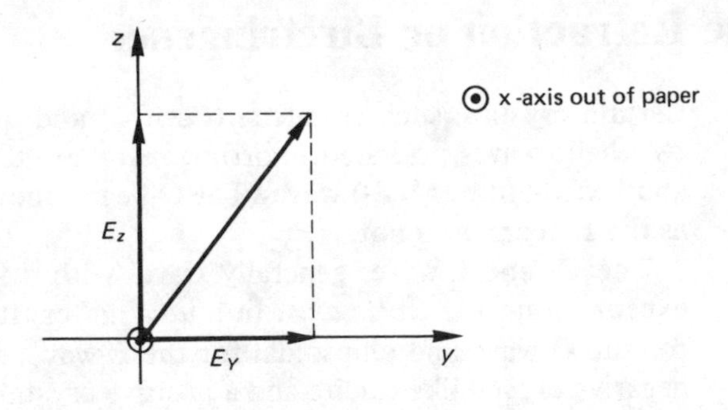

Figure 5.3

As time progresses, the displacements vary and the resultant optical vector traces out a curve in the plane $x = x_o$. This curve gives its name to the overall state of polarisation of the light. In general, it is an ellipse. The relevant equation has the form

$$E_Y^2/A_Y^2 + E_Z^2/A_Z^2 - (2E_YE_Z/A_YA_Z)\cos\delta = \sin^2\delta \qquad (5.4)$$

with the major axis oriented at an angle α to the positive Y-axis, given by

$$\tan 2\alpha = [2A_YA_Z/(A_Y^2 - A_Z^2)]\cos\delta \qquad (5.5)$$

For an observer looking at the light source, i.e. along the negative X-axis, the resultant optical vector rotates *anti-clockwise* (left-handed) around the ellipse if the Z-wave *lags* the Y-wave by an angle δ lying between zero and π. It rotates *clockwise* if δ is given by $\pi < \delta < 2\pi$. Particular cases are shown in Fig. 5.4; each rectangle has sides equal to $2A_Y$ and $2A_Z$. In the special case in which A_Y equals A_Z, circularly polarised light is produced.

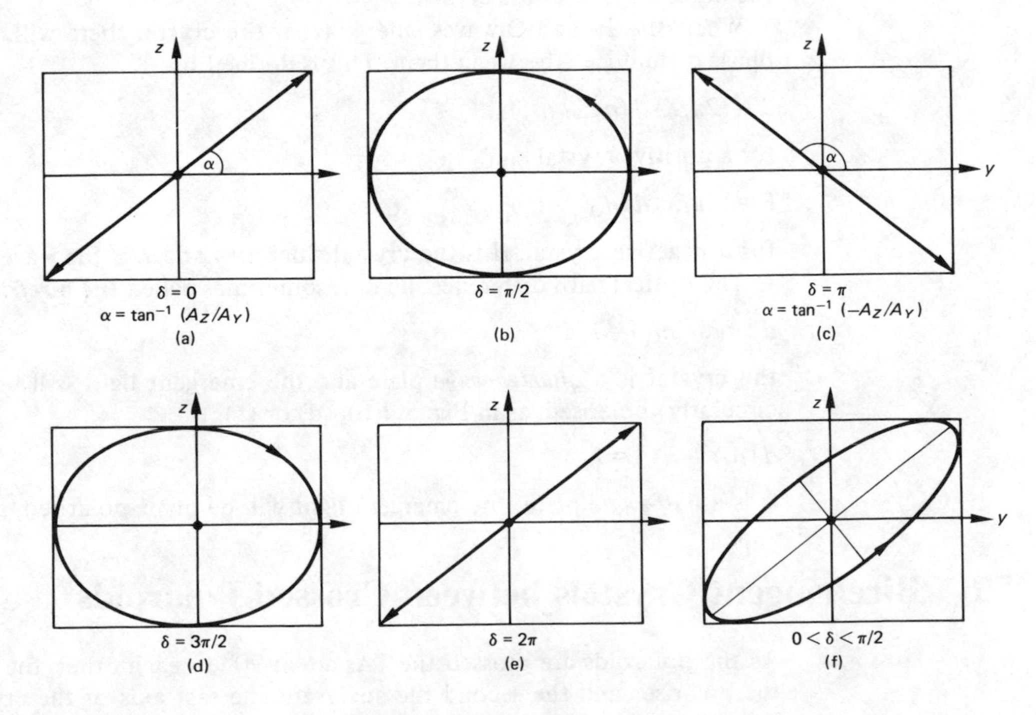

Figure 5.4

5.5 Double Refraction or Birefringence

Certain crystals, such as calcite ($CaCO_3$) and quartz, have the ability to generate two light waves polarised in orthogonal planes. These are called the Ordinary (O-) and Extra-ordinary (E-) waves. The O-beam obeys Snell's laws of refraction whereas the E-beam does not.

The O- and E-waves generally travel with different velocities in these crystals, except along the *optic axis*. In a *uniaxial* crystal the wave surfaces are spherical for the O-wave and ellipsoidal for the E-wave. These are shown in Fig. 5.5 for a negative crystal like calcite and a positive crystal like quartz.

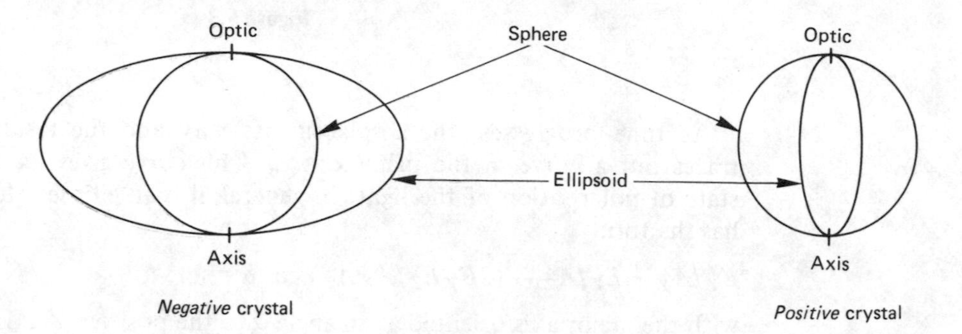

Figure 5.5

Actually, the wave surfaces do not touch in quartz – this is the origin of optical activity in this material. The optic axis can be related to a crystallographic axis of symmetry. In calcite, for example, it is a three-fold axis of rotational symmetry.

The E-wave is polarised in the plane containing the optic axis and the direction of propagation and the O-wave is polarised in the orthogonal plane. The directions of these planes are called the *fast* and *slow* axes, depending upon which wave has the highest speed in the crystal.

When the E- and O-waves emerge from the crystal there will, in general, be a phase difference δ between them. This is defined by

$$\delta = (2\pi/\lambda)\, d\,(n_E - n_O) \tag{5.6}$$

for a positive crystal and

$$\delta = (2\pi/\lambda)\, d\,(n_O - n_E) \tag{5.7}$$

for a negative crystal. d is the crystal thickness and λ is the wavelength *in vacuo*.

The optical path difference, here, is sometimes called the *birefringence*. If

$$d\,|n_E - n_O| = \lambda/4 \tag{5.8}$$

the crystal is a *quarter-wave* plate and the emergent light will be elliptically (or circularly) polarised, as in Fig. 5.4(b), (d) or (f). If

$$d\,|n_O - n_E| = \lambda/2 \tag{5.9}$$

it is a *half-wave* plate, the emergent light will be plane-polarised, as in Fig. 5.4(c).

5.6 Birefringent Crystals between Crossed Polaroids

As the polaroids are crossed the TAs are at 90° to each other; the first polaroid is the polariser and the second the analyser. The fast axis of the crystal is assumed to be along the Z-direction and the optic axis along the Y-direction. The irradiance

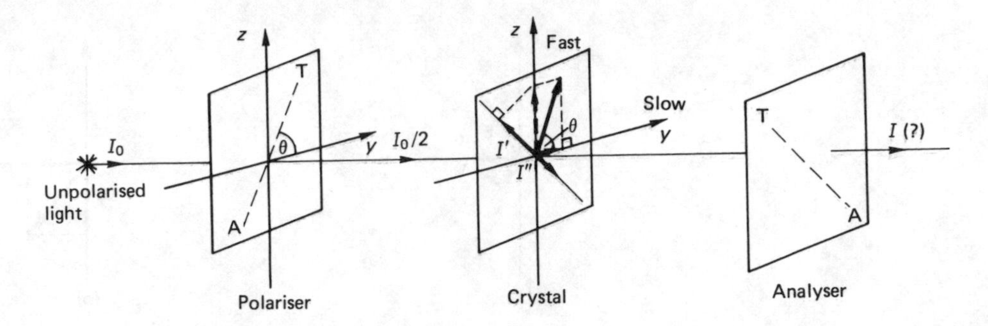

Figure 5.6

of the light incident on the crystal is $I_o/2$ (see Section 5.2). The optical vector of this light is first resolved along the crystal axes to give components equal to $(I_o/2) \cos\theta$ and $(I_o/2) \sin\theta$. Now component irradiances I' and I'' lying parallel to the TA of the analyser may be found. They are both equal to $(I_o/2) \cos\theta \sin\theta$.

As the component waves pass through the crystal a phase difference δ builds up between them. If δ is $2\pi, 4\pi, \ldots, 2\pi m$ (m = integer), the waves will be in phase and the component waves will interfere destructively. No light will be transmitted by the analyser. If $\delta = \pi, 3\pi, \ldots, (2m + 1)\pi$ the waves will interfere constructively to give a resultant irradiance of $I_o \cos\theta \sin\theta$ transmitted by the analyser. If $\delta = \pi/2, 3\pi/2, \ldots, (m + \frac{1}{2})\pi$ either I' or I'', but not both, will lie along the TA of the analyser to give a transmitted irradiance of $(I_o/2) \cos\theta \sin\theta$.

5.7 Vector Notation

It is sometimes useful to express orthogonally polarised beams in vectorial form, viz.

$$\left.\begin{aligned} \mathbf{E}_Y &= \hat{\mathbf{j}} A_Y \cos(kx - \omega t) \\ &\text{and} \\ \mathbf{E}_Z &= \hat{\mathbf{k}} A_Z \cos(kx - \omega t + \delta) \end{aligned}\right\} \quad (5.10)$$

where $\hat{\mathbf{j}}$ and $\hat{\mathbf{k}}$ are unit vectors along the Y-axis and Z-axis, respectively. Then the resultant disturbance is given by

$$\mathbf{E} = \mathbf{E}_Y + \mathbf{E}_Z \quad (5.11)$$

When $\delta = 0$, the waves are in phase, and

$$\mathbf{E} = (\hat{\mathbf{j}} A_Y + \hat{\mathbf{k}} A_Z) \cos(kx - \omega t) \quad (5.12)$$

The amplitude of the resultant wave is constant and equal to $(A_Y^2 + A_Z^2)^{\frac{1}{2}}$; the wave, itself, lies in a plane inclined at an angle $\tan^{-1}(A_Z/A_Y)$ to the positive Y-axis (see Fig. 5.7(a)).

When $\delta = \pi$, $\mathbf{E}_Z$ may be written in the form

$$\mathbf{E}_Z = -\hat{\mathbf{k}} A_Z \cos(kx - \omega t)$$

and the resultant wave by

$$\mathbf{E} = (\hat{\mathbf{j}} A_Y - \hat{\mathbf{k}} A_Z) \cos(kx - \omega t) \quad (5.13)$$

Once again the amplitude has a fixed value. The plane-polarised wave is now inclined at an angle of $\tan(-A_Z/A_Y)$ to the positive Y-axis (see Fig. 5.7(b)).

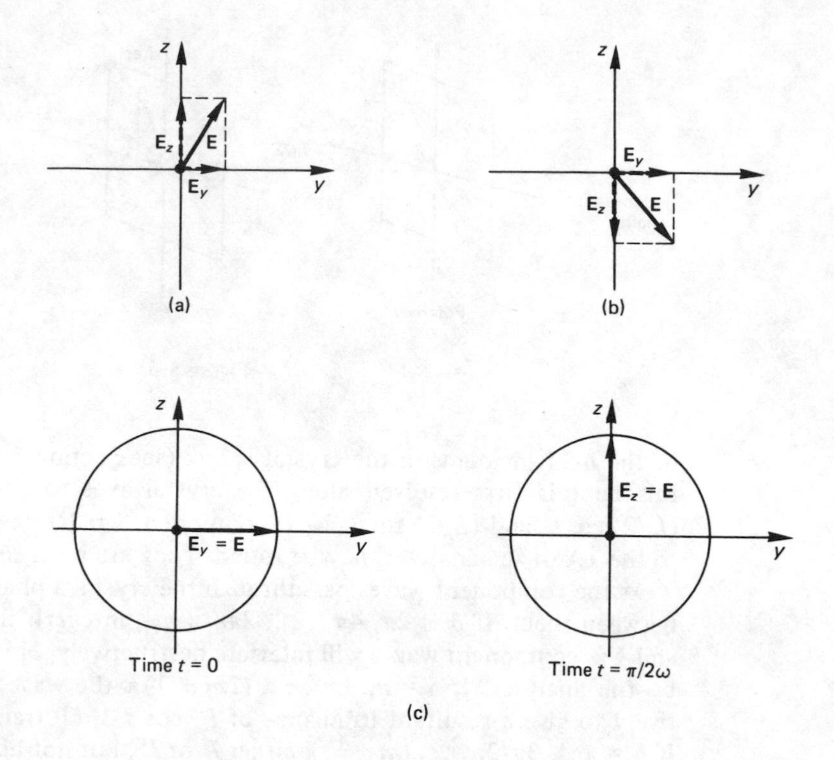

Figure 5.7

Next, consider the case in which $\delta = \pi/2$ and $A_Y = A_Z = A$. Then

$$\left.\begin{aligned} \mathbf{E}_Y &= \hat{\mathbf{j}} A_Y \cos(kx - \omega t) \\ &\text{and} \\ \mathbf{E}_Z &= \hat{\mathbf{k}} A_Z \cos\left(kx - \omega t + \frac{\pi}{2}\right) = -\hat{\mathbf{k}} A_Z \sin(kx - \omega t) \end{aligned}\right\} \quad (5.14)$$

The resultant wave is now given by

$$\mathbf{E} = A\,[\hat{\mathbf{j}} \cos(kx - \omega t) - \hat{\mathbf{k}} \sin(kx - \omega t)] \quad (5.15)$$

E has a constant scalar amplitude A but a direction which varies with time t. In order to determine what is happening to **E** as time proceeds, let us put $x = 0$ simply for convenience sake and find $\mathbf{E}_Y$ and $\mathbf{E}_Z$ at times $t = 0$ and $t = \pi/2\omega$. Figure 7.5(c) shows the result. The E vector has rotated anti-clockwise.

For $\delta = 3\pi/2$:

$$\left.\begin{aligned} \mathbf{E}_Y &= \hat{\mathbf{j}} A_Y \cos(kx - t) \\ &\text{and} \\ \mathbf{E}_Z &= \hat{\mathbf{k}} A_Z \sin(kx - \omega t) \end{aligned}\right\} \quad (5.16)$$

when **E** rotates clockwise.

Following the method outlined in Section 2.7, it is possible to write equations (5.3) in the exponential form as follows

$$\left.\begin{aligned} \mathbf{E}_Y &= \hat{\mathbf{j}} A_Y \exp[\mathrm{i}(kx - \omega t)] \\ &\text{and} \\ \mathbf{E}_Z &= \hat{\mathbf{k}} A_Z \exp[\mathrm{i}(kx - \omega t + \delta)] \end{aligned}\right\} \quad (5.17)$$

Equations (5.3) are the real parts of these expressions.

5.8 Matrix Representation

(a) Jones Vector

The complex form of the wave equations described by equation (5.3) can be expressed in matrix form as

$$\begin{pmatrix} E_Y \\ E_Z \end{pmatrix} = \exp\left[\mathrm{i}\,(kx - \omega t)\right] \begin{pmatrix} A_Y \\ A_Z \exp(\mathrm{i}\delta) \end{pmatrix} \tag{5.18}$$

The pre-multiplier of the right-hand matrix may be omitted because it is multiplied by its complex conjugate in irradiance calculations, when the result is unity. Therefore we may write

$$\begin{pmatrix} E_Y \\ E_Z \end{pmatrix} = \begin{pmatrix} A_Y \\ A_Z \exp(\mathrm{i}\delta) \end{pmatrix} \tag{5.19}$$

The matrix on the right-hand side is called the *Jones vector* (or the *Maxwell column vector*).

For horizontally polarised light, $A_Z = 0$, and the Jones vector is

$$A_Y \begin{pmatrix} 1 \\ 0 \end{pmatrix}$$

For light plane-polarised at 45° to the positive Y-axis, $A_Y = A_Z = A$ and $\delta = 0$, and the Jones vector is

$$A \begin{pmatrix} 1 \\ 1 \end{pmatrix}$$

For clockwise-circularly polarised light, $A_Y = A_Z = A$ and $\delta = 3\pi/2$, and the Jones vector is

$$A \begin{pmatrix} 1 \\ \exp(\mathrm{i}3\pi/2) \end{pmatrix} = A \begin{pmatrix} 1 \\ -\mathrm{i} \end{pmatrix}$$

Any full Jones vector may be reduced to the standard *normalised* form by considering a light beam of unit intensity. These are given in Table 5.1 for some states of polarisation.

(b) Jones Matrix

Let a polarised light beam, represented by its normalised Jones vector e_I^N pass through an optical element and produce a polarised beam e_T^N. Here I stands for 'incident' and T for 'transmitted'. The optical element may change the irradiance level and the state of polarisation of the incident beam. If we assume that this is achieved in a linear way then we can write

$$e_T^N = J e_I^N \tag{5.20}$$

where J is a 2 × 2 matrix, called the *Jones* matrix.

If there is a series of N optical elements with Jones elements J_r, where $r = 1$ to N, then

$$e_{T_1}^N = J_1 e_I^N;\ e_{T_2}^N = J_2 e_{T_1}^N;\ \ldots;\ e_{T_r}^N = J_r e_{T_{r-1}}^N \tag{5.21}$$

Table 5.1 Normalised Jones vectors of some states of polarisation

State of polarisation	Normalised Jones vector
Plane; vertical	$\begin{pmatrix}0\\1\end{pmatrix}$
Plane; horizontal	$\begin{pmatrix}1\\0\end{pmatrix}$
Plane; at 45°	$\frac{1}{\sqrt{2}}\begin{pmatrix}1\\1\end{pmatrix}$
Plane; at 135°	$\frac{1}{\sqrt{2}}\begin{pmatrix}1\\-1\end{pmatrix}$
Circular; right	$\frac{1}{\sqrt{2}}\begin{pmatrix}1\\-i\end{pmatrix}$
Circular; left	$\frac{1}{\sqrt{2}}\begin{pmatrix}1\\i\end{pmatrix}$

Linear optical element	Jones matrix
Polariser; horizontal	$\begin{pmatrix}1 & 0\\0 & 0\end{pmatrix}$
Polariser; vertical	$\begin{pmatrix}0 & 0\\0 & 1\end{pmatrix}$
Polariser; 45°	$\frac{1}{2}\begin{pmatrix}1 & 1\\1 & 1\end{pmatrix}$
Polariser; 135°	$\frac{1}{2}\begin{pmatrix}1 & -1\\-1 & 1\end{pmatrix}$
Quarter-wave plate Fast axis vertical	$\begin{pmatrix}\exp(i\pi/4) & 0\\0 & \exp(-i\pi/4)\end{pmatrix}$
Quarter-wave plate Fast axis horizontal	$\begin{pmatrix}\exp(i\pi/4) & 0\\0 & \exp(i3\pi/4)\end{pmatrix}$

and

$$
\begin{aligned}
e_{T_r}^N &= J_r J_{r-1} \ldots J_2 J_1 e_I^N \\
&= J e_I^N
\end{aligned}
\tag{5.22}
$$

Table 5.2 lists the Jones matrices for various optical elements. Let us take a few cases to illustrate their use.

1. $J = \begin{pmatrix}0 & 0\\0 & 0\end{pmatrix}$

Although this Jones matrix is not listed in Table 5.2 it is worth starting off with it

Table 5.2 Jones matrices for some optical elements

Linear optical element	Jones matrix
Circular polariser; right-handed	$\frac{1}{2}\begin{pmatrix} 1 & i \\ -i & 1 \end{pmatrix}$
Circular polariser; left-handed	$\frac{1}{2}\begin{pmatrix} 1 & -i \\ i & 1 \end{pmatrix}$

because it means that there is no transmitted beam. That is it characterises a complete absorber.

2. $J = \begin{pmatrix} 1 & 0 \\ 0 & 0 \end{pmatrix}$

Only the Y-component wave is transmitted; the optical element completely absorbs the Z-component wave. This is the Jones matrix for a polaroid sheet having its transmission axis parallel to the Y-axis. The corresponding Jones matrix for polaroid with its transmission axis parallel to the Z-axis is also given in Table 5.2.

3. $J = \frac{1}{2}\begin{pmatrix} 1 & 1 \\ 1 & 1 \end{pmatrix}$

Let a horizontally polarised beam be incident on this optical element. Then

$$e_T^N = \frac{1}{2}\begin{pmatrix} 1 & 1 \\ 1 & 1 \end{pmatrix}\begin{pmatrix} 1 \\ 0 \end{pmatrix}$$

$$= \frac{1}{2}\begin{pmatrix} 1 \\ 1 \end{pmatrix}$$

This represents a plane-polarised beam inclined at 45° to the positive Y-axis. So the Jones matrix characterises polaroid with a transmission axis at 45° to the Y-axis. Notice that the irradiance of the transmitted beam is

$$(\tfrac{1}{2})^2 + (\tfrac{1}{2})^2 = \tfrac{1}{2}$$

which is one half of the incident irradiance. You can check this using Malus's law.

4. $J = \begin{pmatrix} \exp(i\pi/4) & 0 \\ 0 & \exp(-i\pi/4) \end{pmatrix}$

For a plane-polarised beam inclined at 45° to the positive Y-axis we have

$$e_T^N = \frac{1}{\sqrt{2}}\begin{pmatrix} \exp(i\pi/4) & 0 \\ 0 & \exp(-i\pi/4) \end{pmatrix}\begin{pmatrix} 1 \\ 1 \end{pmatrix}$$

$$= \frac{1}{\sqrt{2}}\exp(i\pi/4)\begin{pmatrix} 1 & 0 \\ 0 & -i \end{pmatrix}\begin{pmatrix} 1 \\ 1 \end{pmatrix}$$

$$= \frac{1}{\sqrt{2}}\exp(i\pi/4)\begin{pmatrix} 1 \\ -i \end{pmatrix}$$

This means that a right-handed circularly polarised beam is produced.

(c) Cautionary Note

Throughout Sections 5.3 to 5.8, the phase angle of the wave equation has been defined through the cosine term:

$\cos(kx - \omega t + \phi)$

Care has to be taken if other forms of this 'expression' are used, e.g.

$\cos(\omega t - kx + \phi')$, $\sin(kx - \omega t + \phi')$ or $\sin(\omega t - kx + \phi')$

With component waves expressed by

$E_Y = A_Y \cos(\omega t - kx + \phi_1)$

and

$E_Z = A_Z \cos(\omega t - kx + \phi_2)$

the Z-component wave *leads* the Y-component wave by δ $(= (\phi_2 - \phi_1))$ and the opposite results to Section 5.3 apply. Now, δ given by $0 < \delta < \pi$ means that clockwise elliptically polarised light is obtained, in general. It is advisable to use only one form of the wave equations in order to avoid any confusion which might otherwise arise.

5.9 Worked Examples

Example 5.1

Two coherent plane-polarised waves of amplitudes A and $A\sqrt{2}$ travel along the X-axis of a right-handed Cartesian coordinate system. The plane of polarisation of the first wave is parallel to the Y-axis and that of the second lies at an angle of $\pi/4$ with respect to the Y-axis. The second wave has a phase delay of $\pi/2$ over the first wave. Write equations describing the resultant wave and discuss its state of polarisation.

To find the resultant wave the Principle of Superposition must be applied to waves polarised in planes at right angles – in this case, the X–Y and X–Z planes. This means that the displacement of the second wave must be resolved into two components; one along the Y-axis and the other along the Z-axis. The total displace-

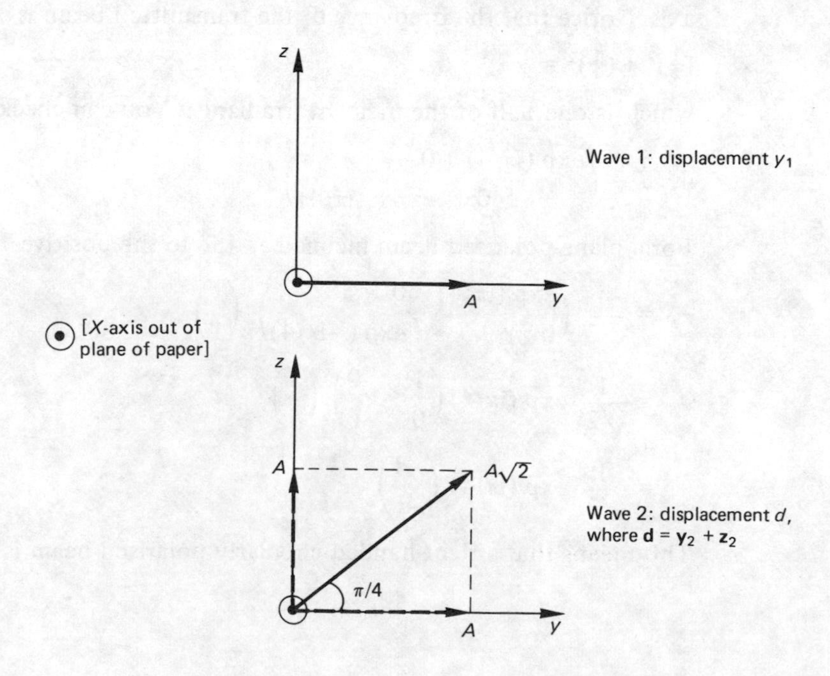

ment along the Y-axis can then be determined taking into account the phase delay. The equation of the wave polarised in the X–Y plane may now be compared with the equation of the wave polarised in the X–Z plane, and the state of polarisation established.

Data	*Given:*	*Wave 1*	*Wave 2*
		amplitude $= A$	amplitude $= A\sqrt{2}$
			$\delta = \pi/2$
		parallel to Y-axis	lies at angle of $\pi/4$ to Y-axis
	Unknown:	resultant displacement $\mathbf{y} = \mathbf{y}_1 + \mathbf{d}$	
		state of polarisation	
Relevant equations		$y_1 = A\cos(kx - \omega t)$	
		$d = A\sqrt{2}\cos(kx - \omega t + \frac{1}{2}\pi)$	

Resolve displacement d for wave 2 along the Y- and Z-axes to give

$$y_2 = A\cos\left(kx - \omega t + \frac{\pi}{2}\right) \tag{1}$$

and

$$z_2 = A\cos\left(kx - \omega t + \frac{\pi}{2}\right) \tag{2}$$

The total displacement along the Y-axis is

$$y = y_1 + y_2$$

$$= 2A\cos\left(kx - \omega t + \frac{\pi}{4}\right)\cos\frac{\pi}{4} \tag{3}$$

$$= A\sqrt{2}\cos\left(kx - \omega t + \frac{\pi}{4}\right)$$

(2) and (3) describe the resultant wave motion. By inspection, we see that the amplitudes are different and the Z-wave lags the Y-wave by $\pi/4$. Therefore the resultant wave is left-handed elliptically polarised.

NOTE

The initial phase of wave 1 is not stated. It is assumed to be zero.

Example 5.2

The irradiances of two unpolarised light sources are compared using a polariscope. It is found that the transmitted irradiances are the same if the transmission axes of the polariser and analyser are at an angle of $\pi/4$ to each other with source 1 and at an angle of $\pi/6$ with source 2. Find the ratio between the irradiances of the sources.

Malus's law relates the transmitted irradiance I_T of a polaroid sheet to the incident irradiance I_o by taking into account the angle θ between the transmission axis and

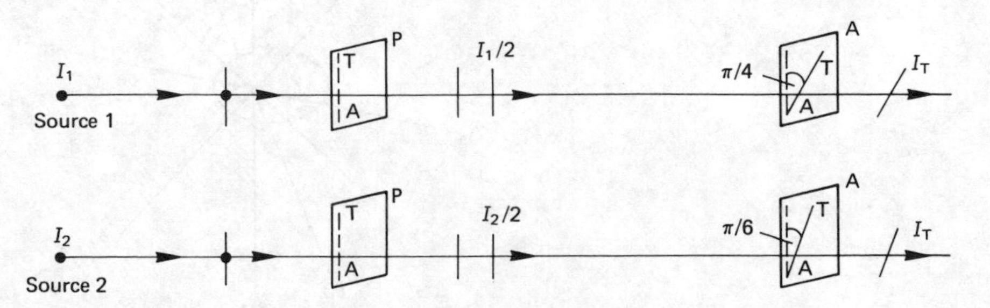

the optical vector of the incident light. In order to proceed, the random fluctuations in the plane of polarisation of the light from the unpolarised sources are resolved into components lying in the plane of incidence and at right angles to it. For this example, Malus's law needs to be applied at the analyser.

Data		$I_T = I_o \cos^2 \theta$
Relevant equation	*Given:*	angle θ between axes of P and A for source $1 = \pi/4$
		angle θ between axes of P and A for source $2 = \pi/6$
		I_T is the same for both sources
	Unknown:	I_1/I_2

$$\left.\begin{array}{l}(I_T)_{P_1} = I_1/2 \quad \text{for source 1}\\(I_T)_{P_2} = I_2/2 \quad \text{for source 2}\end{array}\right\} \tag{1}$$

Therefore, after transmission by the analyser

$$I_T = (I_1/2)\cos^2 \frac{\pi}{4} \quad \text{for source 1} \tag{2}$$

and

$$I_T = (I_2/2)\cos^2 \frac{\pi}{6} \quad \text{for source 2} \tag{3}$$

Hence

$$I_1/2 = 3I_2/4 \tag{4}$$

or

$$I_1/I_2 = 3/2$$

NOTE

For convenience, the plane of incidence is taken to be the plane of the paper.

Example 5.3

A very large number $(N + 1)$ of polaroids are arranged in a sandwich. The transmission axis of each polaroid is at a constant angle α with respect to its predecessor. Thus the last polaroid is at an angle of $(\phi = N\alpha)$ from the first. Suppose that plane-polarised light of irradiance I_o is incident on the first polaroid which has its transmission axis parallel to the plane of polarisation of the light. Determine the output irradiance from the sandwich.

Here again, Malus's law is applied at each polaroid, taking into account the angle between the transmission axis and the optical vector of the incident light. A general expression for the output irradiance can be obtained by induction.

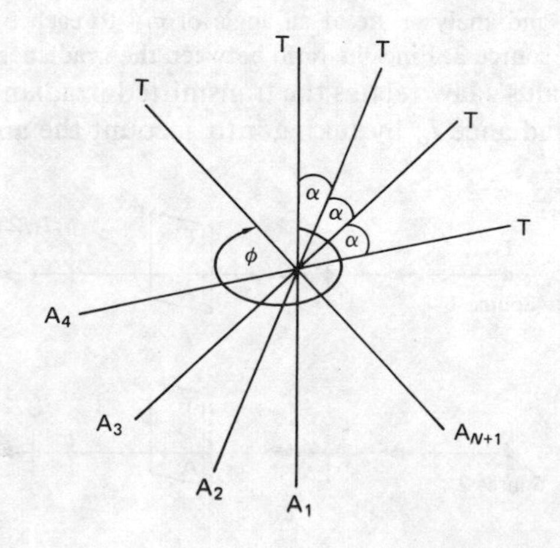

Data	*Given:*	number of polaroids = $N + 1$
		angle between transmission axes = α
		angle between first and last transmission axes = $\phi = N\alpha$
		irradiance of incident light = I_o
	Unknown:	output irradiance = $(I_T)_{N+1}$
Relevant equation		$I_T = I_o \cos^2 \theta$

The irradiance transmitted by polaroid 1 is I_o, therefore

$$(I_T)_2 = I_o \cos^2 \alpha \tag{1}$$

and

$$(I_T)_3 = I_o \cos^2 \alpha \cdot \cos^2 \alpha = I_o \cos^4 \alpha \tag{2}$$

Following on in this way we have

$$(I_T)_4 = I_o \cos^6 \alpha$$

and

$$(I_T)_5 = I_o \cos^8 \alpha$$

We are now in a position to generalise for the Nth polaroid, when

$$(I_T)_N = I_o \cos^{2N-2} \alpha \tag{3}$$

and the emergent irradiance is given by

$$(I_T)_{N+1} = I_o \cos^{2N-2} \alpha \cdot \cos^2 \alpha = I_o \cos^{2N} \alpha = I_o \cos^{2N} (\phi/N) \tag{4}$$

As N is large, use the Taylor expansion to give

$$\cos^{2N} (\phi/N) = (1 - \phi^2/N^2 + \ldots +)^{2N} = (1 - 2\phi^2/N + \ldots +)$$

Therefore the output irradiance is smaller than the incident irradiance by the amount $2I_o\phi^2/N$.

NOTE

The transmitted irradiance will still be almost I_o even if the first and last polaroids are crossed. This is because $2\phi^2 I_o/N$ tends to zero. This is not the case with a *pair* of crossed polaroids.

Example 5.4

Light reflected from the surface of a calm lake is found to be completely plane-polarised. Estimate the sun's angular position above the surface of the lake (take $n_w = 1.330$).

When the angle of incidence of the sunlight on the surface of the calm water equals the Brewster (or polarising) angle the plane-polarised component does not appear in the reflected light. This occurs because the optical vector of the reflected p-light would need to vibrate along the direction of propagation – this is inadmissible because light has a transverse waveform.

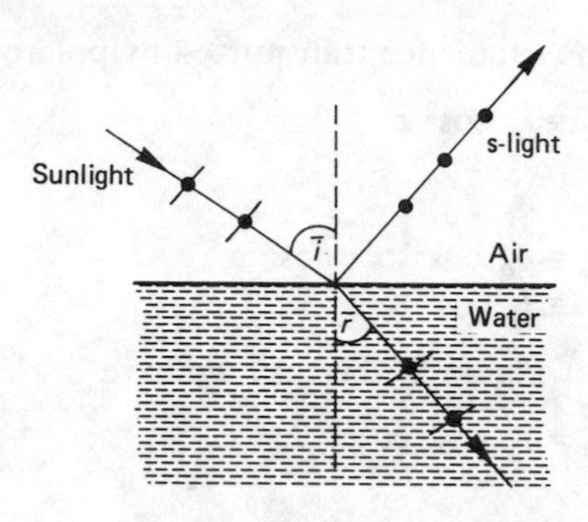

Data	*Given:*	$n_w = 1.330$
	Unknown:	polarising angle $\bar{i}$
Relevant equations		$n_w = \tan \bar{i}$
		$n_w = \sin i / \sin r$

Now

$$\bar{i} = \tan^{-1} n_w$$

Therefore the sun's angular position above the horizontal is equal to $90° - \tan^{-1} n_w$

so

$$\bar{i} = \tan^{-1} 1.330 = 58.00°$$

and

$$\text{angular position} = 90° - 58.00° = 32.00°$$

NOTES

(i) It is assumed that the sunlight is unpolarised. This is not strictly true because sunlight will be scattered by the atmosphere; this is the origin of the blue colour of the sky.

(ii) As the refractive index is dependent on wavelength, this calculation is correct for one wavelength only.

(iii) As mentioned in the Notes to Example 1.4 a refractive index of 1.330 is characteristic of pure water. Lake water will almost certainly contain dissolved ions and other impurities which will increase the refractive index somewhat.

Example 5.5

Unpolarised light is incident on a water surface at the polarising angle. The refracted ray meets a block of glass ($n_g = \frac{3}{2}$) immersed in the water ($n_w = \frac{4}{3}$). The flat upper surface of the block makes an angle α with the water surface. Calculate α if the light reflected from the block is also plane-polarised.

Brewster's law must be applied at both the air/water and the water/glass interfaces. The correct value of the refractive index of glass immersed in water must be used for the latter calculation.

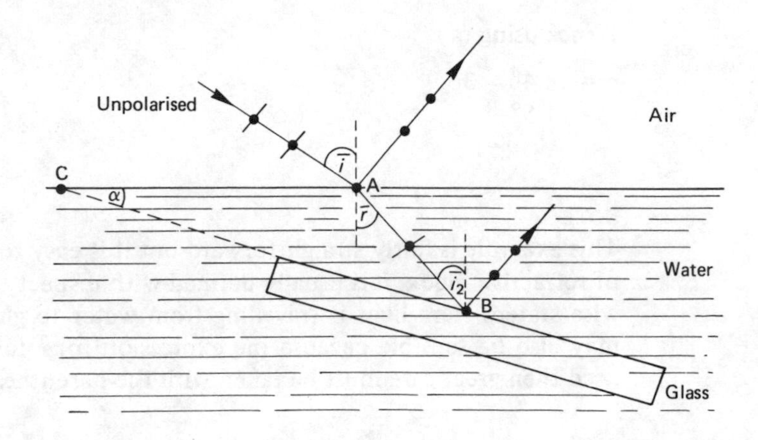

Data	*Given:*	${}_An_w = 4/3$, ${}_An_g = 3/2$ } with respect to air
	Unknown:	$\bar{i}_1$
		$\bar{r}$
		$\bar{i}_2$
		${}_wn_g$
		α
Relevant equations		${}_1n_2 = \tan \bar{i}$
		$= \sin i/\sin r$

At air/water interface:

$\tan \bar{i}_1 = {}_An_w$

$\therefore \bar{i}_1 = \tan^{-1} {}_An_w$ (1)

As

${}_An_w = \sin \bar{i}_1/\sin \bar{r}$

$\bar{r} = \sin^{-1} [[\sin(\tan^{-1} {}_An_w)]/{}_An_w]$ (2)

Also, at water/glass interface:

$\tan \bar{i}_2 = {}_wn_g$

where

${}_wn_g = {}_An_g/{}_An_w$ (3)

In triangle ABC,

$\alpha + 90 + \bar{r} + 90 - \bar{i}_2 = 180$

from which

$\alpha = \bar{i}_2 - \bar{r}$ (4)

Now

$\bar{r} = \sin^{-1}$ [[sin ($\tan^{-1}$ 4/3)]/(4/3)]
$= \sin^{-1}$ [sin 58.0°/(4/3)]
= 36.9°

and

$_w n_g = (3/2)/(4/3) = 9/8$
$\therefore \bar{i}_2 = \tan^{-1}$ (9/8)
= 48.4°

Hence using (4)

$\alpha = 48.4° - 36.9°$
$= 11.5°$

NOTE

This example is fairly straightforward but it is easy to forget the correct definition of refractive index; it is usually defined with respect to air or vacuum unless otherwise stated. Here light is travelling from water to glass. Step-by-step substitution may also be sensible because the expression for $\bar{r}$ looks a little formidable; if it is used then great care must be taken with the parentheses.

Example 5.6

Mica cleaves naturally along planes which are perpendicular to the *X*-axis of a right-handed Cartesian coordinate system. The refractive indices characterising the two component waves which travel along this direction are 1.5692 and 1.6049. Determine the metrical thickness of a quarter-wave plate for a wavelength of 600 nm and normal incidence.

Mica is doubly refracting (or birefringent). This means that when unpolarised light is incident normally on a mica slice, cut so that its optic axis lies parallel to the surface, an ordinary (O-) and extra-ordinary (E-) wave is produced, polarised in planes at right angles to each other. These waves travel through the mica at different speeds, v_O and v_E, which means that the O- and E-waves have different refractive indices. There is, therefore, an optical path difference between them when they emerge. If this equals $\lambda/4$ then the mica is called a quarter-wave plate. In this example v_E is higher than v_O.

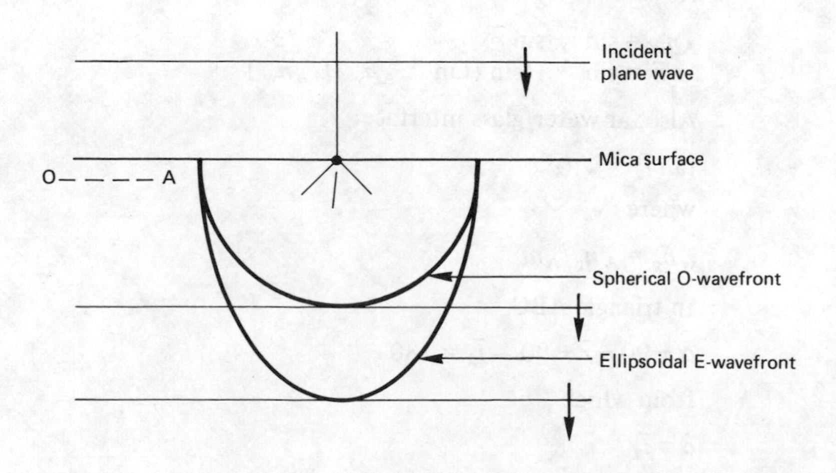

Data	*Given:*	$n_E = 1.5692$
		$n_O = 1.6049$
		optical path difference = $\lambda/4$ = 150 nm
		angle of incidence = 0°
	Unknown:	metrical thickness d
Relevant equations		optical path difference = $d(n_O - n_E) = \lambda/4$
		$v_E = c/n_E$
		$v_O = c/n_O$

Since

$$d = \lambda/4\,(n_O - n_E)$$

we have

$$d = 600/(4 \times 0.0357) \text{ nm}$$
$$= 4200 \text{ nm}$$
$$= 4.20\ \mu\text{m}$$

NOTES

(i) The refractive indices for the O- and E-waves are always determined for a direction normal to the optic axis.

(ii) It is not really essential to know which refractive index characterises the component waves because only the magnitude of their difference is required.

Example 5.7

Devise an optical system for producing a beam of left-handed elliptically polarised light. The major axis of the ellipse is vertical and the ratio of the lengths of the major and minor axes is 2:1.

When a beam of plane-polarised light enters a birefringent crystal the state of polarisation of the emergent light depends on the crystal thickness and the orientation of the optical vector of the incident light relative to the crystal axes. In general, the emergent light is elliptically polarised if the phase difference δ between the O- and E-component waves is given by $0 < \delta < \pi$ or $\pi < \delta < 2\pi$. Circularly polarised light is a special case of elliptical polarisation: it occurs when the amplitudes of the O- and E-waves are equal and δ is $\pi/2$ or $3\pi/2$. The direction of rotation of the resultant optical vector depends on the nature of the crystal, i.e. whether it is positive or negative.

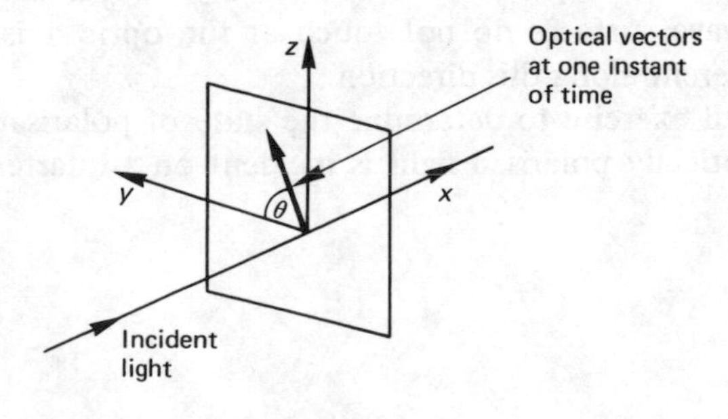

x y z form a right-handed Cartesian coordinate system

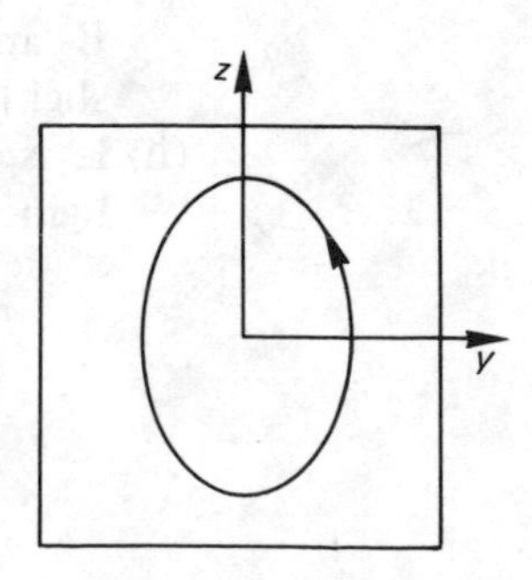

State of polarisation of the emergent light (*x* out of paper)

Data	*Given:*	major axis:minor axis = 2:1
		direction of rotation of optical vector = left-handed
	Unknown:	θ (or amplitudes of E- and O-waves)
		crystal thickness d
		n_E
		n_O
		λ
		type of crystal
		direction of optic axis
Relevant equations		$\delta = 2\pi d \mid (n_E - n_O) \mid / \lambda$
		$\tan \phi = A_Z / A_Y$
		$v_E = c/n_E$
		$v_O = c/n_O$

For the major and minor axes of the ellipse to lie along the Z- and Y-axes δ must be $\pi/2$. Therefore

$$\pi/2 = 2\pi d \mid (n_E - n_O) \mid / \lambda$$

or

$$d \mid n_E - n_O \mid = \lambda/4$$

In other words a quarter-wave plate is required.

As

$$A_Z/A_Y = 2$$
$$\theta = \tan^{-1} 2$$
$$= 63.4°$$

This means that the plane of polarisation of the incident light must be at an angle of 63.4° relative to the positive Y-axis.

The E-wave vibrates in the principal plane, i.e. the plane containing the optic axis and the direction of propagation. Therefore, with a positive crystal like quartz, if the optic axis is parallel to the Z-axis of the crystal the E-wave will travel more slowly through the crystal than the O-wave, which vibrates parallel to the Y-axis. This is the requirement for left-handed elliptical polarisation.

With a negative crystal, like calcite, the optic axis should be parallel to the Y-axis.

NOTES

(i) Although quartz is a positive birefringent crystal it is also optically active; the E- and O-wave surfaces do not touch at the optic axis, i.e. v_E and v_O are slightly different along this direction.

(ii) It is a useful exercise to determine the state of polarisation of the emergent light if elliptically polarised light is incident on a quarter- or half-wave plate.

Example 5.8

Describe the state of polarisation of the waves represented by

(a) $E_Y = A \cos(kx - \omega t)$ $\quad E_Z = A \sin(kx - \omega t)$
(b) $E_Y = A \sin(kx - \omega t)$ $\quad E_Z = A\sqrt{3} \sin(kx - \omega t)$
(c) $E_Y = A \cos(kx - \omega t)$ $\quad E_Z = 2A \cos(kx - \omega t - \pi/4)$

In order to obtain the state of polarisation, attention must be paid to the phase difference between the component waves and the individual amplitudes. The former will enable the direction of rotation of the resultant optical vector to be found as well as a general answer to the state of polarisation. The information obtained from the amplitudes will allow a specific answer to be obtained.

(a) Writing E_Z as

$$E_Z = A \cos\left(kx - \omega t - \frac{\pi}{2}\right)$$

shows that the Z-wave leads the Y-wave by $\pi/2$. Thus, following the rule given in Section 5.4, the state of polarisation is in general right-handed elliptical polarisation with major and minor axes parallel to the Y- and Z-axes of a right-handed Cartesian coordinate system. However, as the amplitudes are equal the actual state of polarisation is right-handed circular polarisation.

(b) The component waves are completely in phase. Therefore we are dealing with plane-polarised light. The individual amplitudes allow us to determine the inclination of the plane of vibration with the positive Y-axis as

$$\tan \alpha = A\sqrt{3}/A = \sqrt{3}$$

i.e.

$$\alpha = 60°$$

Hence

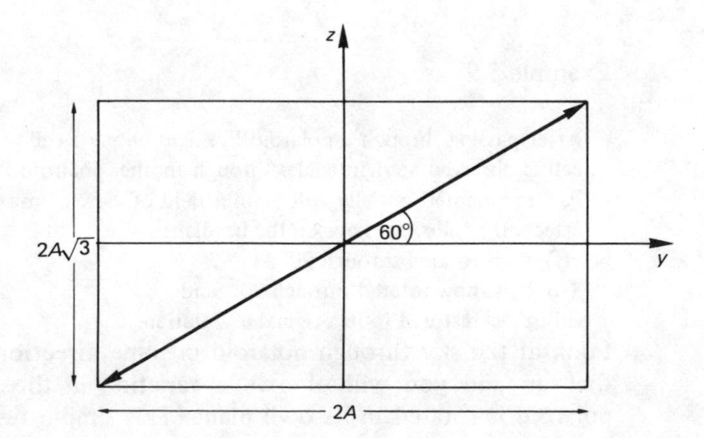

(c) Here, the Z-component wave leads the Y-component wave by $\pi/4$. This information immediately tells us that the resultant wave is in a right-handed state. The amplitudes are unequal, which means that it is right-handed elliptically polarised.

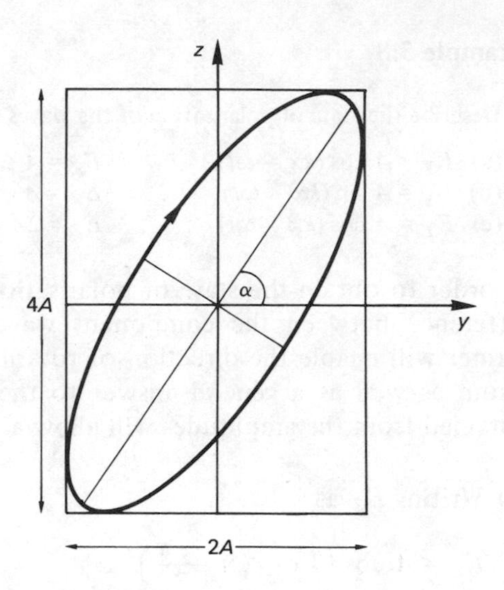

NOTE

The angle α, which the major axis makes with the positive Y-axis, can be found using

$$\tan 2\alpha = [2A_Y A_Z/(A_Z^2 - A_Y^2)] \cos \delta$$

This gives

$$2\alpha = \tan^{-1}\left[\left(4A^2/A^2\right) \cos \frac{\pi}{4}\right] = \tan^{-1} 2\sqrt{2} = 70.5°$$

and

$$\alpha = 35.25°$$

Example 5.9

> After passing through a polaroid P_1, light enters a cell containing a scattering medium. This cell is observed at right angles through another polaroid P_2. The transmission axes of P_1 and P_2 are oriented initially to obtain a field of view of maximum irradiance. Determine what effects the following have on the irradiance:
> (i) P_1 is rotated through 90°;
> (ii) P_2 is now rotated through 90°; and
> (iii) P_1 is restored to its original orientation.

Look at the sky through polaroid in some direction away from the direct line of the sun and you will observe a variation in the transmitted irradiance as the polaroid is rotated in its own plane. This simple test shows that light is scattered by the atmosphere towards your eye and, what is more, is plane-polarised, or, more accurately, has a plane-polarised component. That blue light is scattered more than red is obvious! Scattering is wavelength dependent: it obeys a λ^{-4} law.

Scattering can be simulated on a much smaller scale by directing light into the smoke from a lighted cigarette. Light scattered sideways by the smoke particles has a bluish hue and is plane-polarised. If plane-polarised light is incident on the scattering medium, then there may well be certain directions along which no light is observed. This occurs if the optical vector of the incident light is parallel to this direction. Remember that light is a transverse wave motion.

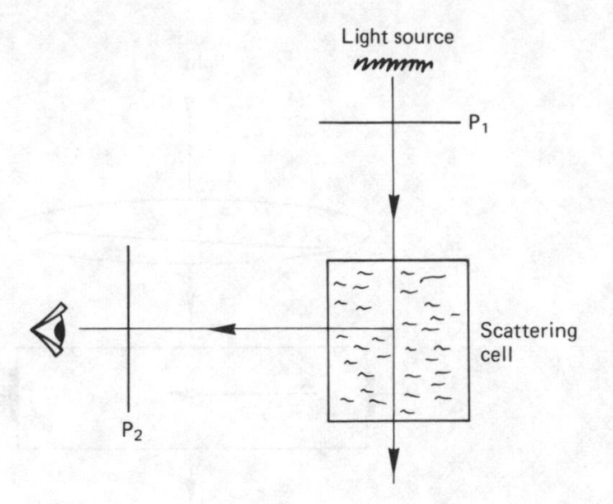

Data	*Given:*	transmission axes of P_1 and P_2 are parallel
	Unknown:	direction of the optical vector of the light transmitted by P_1

(i) Scattered light will not be observed if the transmission axis of P_1 lies in the plane of the paper. We will therefore assume that the transmission axis of P_1 is at right-angles to the plane of the paper, which means that the transmission axis of P_2 must also lie in this direction for the transmitted light to have maximum irradiance. Now if P_1 is rotated through 90° we are back to the case just described, and *no* transmitted light will be produced.

(ii) Rotation of P_2 through 90° will not alter this situation because no scattered light is directed towards it.

(iii) Restoring P_1 to its initial position will allow light to be scattered towards P_2 but as the latter's transmission axis is crossed with that of P_1 *no* transmitted light will be observed.

NOTE

Although the irradiance level of the scattered light varies with λ^{-4}, the conclusions drawn here are universally applicable.

Example 5.10

A Kerr cell is placed between crossed polaroids. It is filled with water having a Kerr constant of 5.23×10^{-14} V^{-2} m. The electric field is increased gradually until the light transmitted by the analyser reaches a maximum irradiance. If the value of this electric field is 7.4×10^4 V cm^{-1}, determine the length of water that the light passes through.

Many normally isotropic substances behave like uniaxial crystals (usually positive) when placed in an electric field. The optic axis lies along the direction of the electric field and becomes an axis of symmetry for the substance.

If the principal refractive indices for light vibrating parallel to the electric field E and perpendicular to it, respectively, are $n_\parallel$ and $n_\perp$, then the birefringence is given by

$$L\,(n_\parallel - n_\perp) = K\lambda E^2 L$$

where K is the Kerr constant and L is the geometrical path length. Light will be completely transmitted by the analyser if the plane of polarisation is parallel to its transmission axis. This will occur if the O- and E-waves are out of phase by π on

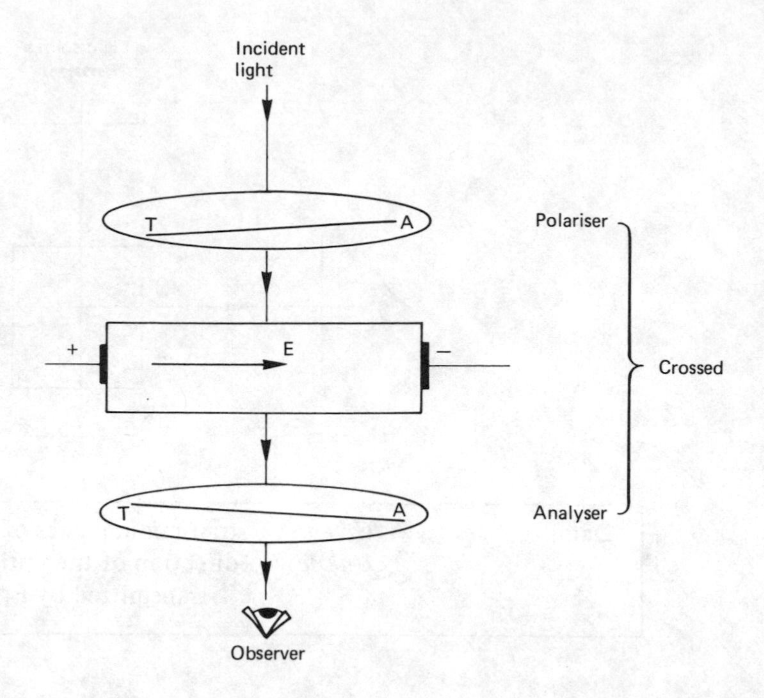

emerging from the Kerr cell. So if L is the length of water traversed by the light we have

$$(2\pi/\lambda)\, L\, (n_{\parallel} - n_{\perp}) = \pi$$

Data	*Given:*	$K = 5.23 \times 10^{-14}\ \mathrm{V^{-2}\ m}$
		$E = 7.1 \times 10^{6}\ \mathrm{V\ m^{-1}}$
	Unknown:	L
		λ
		$n_{\parallel}$
		$n_{\perp}$
Relevant equation		$\mathrm{OPD} = L\,(n_{\parallel} - n_{\perp}) = K\lambda E^2 L$

Using

$$(2\pi/\lambda)\, L\, (n_{\parallel} - n_{\perp}) = \pi$$

we find that

$$2\pi K E^2 L = \pi$$

or

$$\begin{aligned} L &= 1/2KE^2 \\ &= 1/[2 \times 5.23 \times 10^{-14} \times (7.1 \times 10^{6})^2] \\ &= 0.19\ \mathrm{m} \\ &= 19\ \mathrm{cm} \end{aligned}$$

NOTES

(i) This is an example of non-linear optics; the birefringence depends on the square of the electric field strength. For this reason the Kerr effect is sometimes called the *quadratic* electro-optic effect.

(ii) The Kerr constant depends on wavelength and temperature. The value quoted is for yellow sodium light (589.3 nm) and 20°C.

(iii) The water molecule is polarised to some extent; the oxygen atom has a partial negative charge and the hydrogen atoms have partial positive charges.

The electric field aligns the molecules so that they 'stretch' from one electrode to the other. There is, therefore, a delay time before the effect is observed.

(iv) Birefringence can also be induced by a magnetic field. Then it is known as the *Cotton–Mouton* effect.

Example 5.11

A sugar solution of unknown concentration is placed in a tube, 30.0 cm long, and examined in a polarimeter. It is found that the plane of polarisation of the light used is rotated through 35.0°. Determine the concentration of the sugar solution given that the specific rotation is 66.5°.

Section 5.5 indicated that the O- and E-wave surfaces for quartz do not touch along the optic axis direction, and that this is the origin of optical activity in quartz. What this means is that plane-polarised light sent along the optic axis has its plane of polarisation rotated through an angle which depends on the length of quartz it travels through. The amount of rotation also depends on the wavelength; there is a λ^{-2} dependence.

Many other substances exhibit this effect, e.g. turpentine, sodium chlorate, sugar crystals and solution. The rotation may be to the left or to the right. Substances producing the former are laevo-rotatory and the latter are dextro-rotatory. Optical activity is exhibited by those substances which do not possess a plane of symmetry.

The specific rotation of a liquid is defined by

$$SR = \theta / c\,L$$

where θ is the measured rotation, L is the length of the liquid column in decimetres and c is the concentration in grammes of solute per cm^3 of solvent.

The polarimeter is an intrument for accurately measuring the angle of rotation of the plane-polarised light. It usually incorporates some device for splitting the field of view into two halves, called a *half-shadow device* – the Laurent half-wave plate is one example.

Data	*Given:*	$L = 3.00$ dm
		$\theta = 35.0°$
		SR = 65.5°
	Unknown: c	
Relevant equation	$SR = \theta / Lc$	

Direct substitution gives

$$c = \theta/(L \times SR)$$
$$= 35.0/(65.5 \times 3.00) \text{ g cm}^{-3}$$
$$= 0.179 \text{ g cm}^{-3}$$

NOTES

(i) In this example the quoted specific rotation is for sodium yellow light.

(ii) It is possible for the angle of rotation to be greater than 360°. In such cases it is necessary to experiment with tubes of different lengths.

(iii) Natural sugar is D-active. The country of origin is irrelevant, as is the type of sugar plant, i.e. sugar cane or sugar beet.

(iv) All but one of the amino acids are L-active; glycine is the odd man out – it is the simplest. Proteins, which are comprised of groups of amino acids, produce L-active 'bits' when they are broken up.

Example 5.12

A sheet of polaroid is placed immediately after the primary light source of a Young's slits arrangement. Before each slit is placed a quarter-wave plate; the fast axis of one is parallel to the slow axis of the other. Discuss the appearance of the fringe system as the polaroid is rotated in its own plane.

The polaroid sheet allows plane polarised light to be incident on the quarter-wave plates. These introduce a phase delay of $\pi/2$ between the O- and E-waves. The transmitted light will, therefore, be elliptically polarised with major and minor axes aligned with the plate axes. The direction of rotation of the optical vector will, however, be different for the two plates. Only when the transmission axis of the polaroid is at angle of $\pi/4$ to the plate axes will the light incident on the slits be circularly polarised.

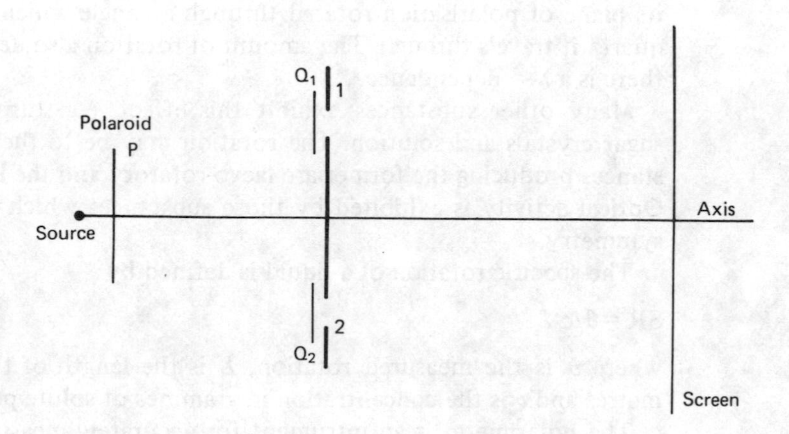

With the transmission axis of the polaroid parallel to either plate axis the light incident on the slits has the same plane of polarisation. Therefore, fringes will be observed. As the transmission axis varies its direction elliptically polarised light will be incident on the slits. As the degree of plane-polarisation is greatest along the major axis fringes will again be observed. They will not be as intense as previously. Lastly, when the transmission axis is at $\pi/4$ to the plate axes, circularly polarised light is generated. In this case there is no preferred direction of polarisation and no fringes are obtained; the screen is uniformly illuminated.

Example 5.13

Describe the state of polarisation of the wave represented by:

$$\mathbf{E}_Y = \mathbf{\hat{j}} A \cos (kx - \omega t)$$

$$\mathbf{E}_Z = \mathbf{\hat{k}} A \cos \left(kx - \omega t - \frac{\pi}{4} \right)$$

The state of polarisation can be determined explicitly by calculating the resultant optical vector E in the plane $x = 0$ at several times – three values should be sufficient, e.g. $t = 0, \pi/4\omega$ and $\pi/2\omega$. A graphical drawing can then be constructed which will indicate the direction of rotation of E. By inspection, as δ is not $\pm \pi/2$, the wave must be elliptically polarised.

Data	*Given:*	$\mathbf{E}_Y$
		$\mathbf{E}_Z$
		equal amplitudes A
		$\delta = -\pi/4$
	Unknown:	direction of rotation of $\mathbf{E}$
		orientation α of ellipse relative to positive Y-axis
Relevant equation		$\tan 2\alpha = 2A_Y A_Z/(A_Y^2 - A_Z^2)$

At time $t = 0$

$$\mathbf{E} = \hat{\mathbf{j}}A + \hat{\mathbf{k}}A \cos\left(\frac{-\pi}{4}\right)$$

$$= \hat{\mathbf{j}}A + \frac{\hat{\mathbf{k}}A}{\sqrt{2}} \qquad (1)$$

At time $t = \pi/4\omega$

$$\mathbf{E} = \hat{\mathbf{j}}A \cos\left(\frac{-\pi}{4}\right) + \hat{\mathbf{k}}A \cos\left(\frac{-\pi}{2}\right)$$

$$= \frac{\hat{\mathbf{j}}A}{\sqrt{2}} \qquad (2)$$

At time $t = \pi/2\omega$

$$\mathbf{E} = \hat{\mathbf{j}}A \cos\left(\frac{-\pi}{2}\right) + \hat{\mathbf{k}}A \cos\left(\frac{-3\pi}{2}\right)$$

$$= \frac{-\hat{\mathbf{k}}A}{2} \qquad (3)$$

Next construct a phasor diagram, similar to Fig. 5.3. The general shape of the ellipse can be found by joining up the tips of the vectors, as shown. The ellipse is right handed.

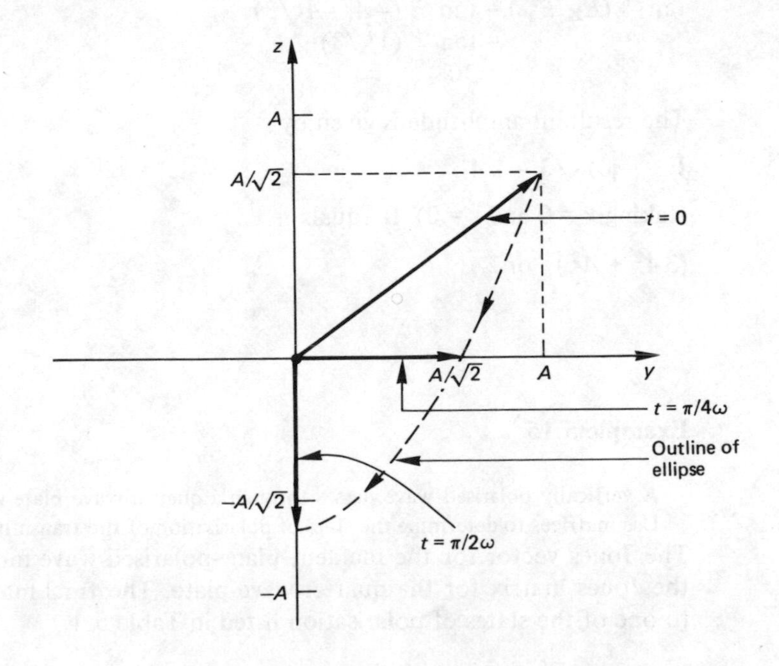

Also

$$\tan 2\alpha = [2A^2/(A^2 - A^2)] \cos \frac{\pi}{4} \quad (4)$$
$$= \infty$$

giving

$2\alpha = \pi/2$

and

$\alpha = \pi/4$

So the major axis is inclined at an angle of 45° to the positive Y-axis.

Example 5.14

Describe the state of polarisation of the wave

$\mathbf{E}_Y = -\hat{\mathbf{j}}A\sqrt{3}\cos(kx - \omega t)$
$\mathbf{E}_Z = -\hat{\mathbf{k}}A\cos(kx - \omega t)$

Look at the phase angle and decide whether the wave components are in phase. If this is the case, as here, then the ratio of the amplitudes will allow the orientation of the plane of vibration to be found.

Data	*Given:*	$\mathbf{E}_Y$
		$\mathbf{E}_Z$
		$\delta = 0$
	Unknown:	E
		orientation
Relevant equation		$E = E_Y + E_Z$

The resultant wave is plane-polarised. It is inclined at an angle, given by

$$\tan^{-1}(E_Z/E_Y) = \tan^{-1}(-A/-A\sqrt{3}) \quad (1)$$
$$= \tan^{-1}(1/\sqrt{3})$$
$$= 30°$$

The resultant amplitude is given by

$$\mathbf{E} = -\hat{\mathbf{j}}A\sqrt{3} - \hat{\mathbf{k}}A \quad (2)$$

(taking $x = 0$ and $t = 0$). It equals

$(3A^2 + A^2)^{\frac{1}{2}}$ or $2A$.

Example 5.15

A vertically polarised wave passes through a quarter-wave plate with its fast axis horizontal. Use matrices to determine the state of polarisation of the transmitted wave.

The Jones vector for the incident plane-polarised wave must be pre-multiplied by the Jones matrix for the quarter-wave plate. The final matrix can then be related to one of the states of polarisation listed in Table 5.1.

Data	*Given:*	$e_I^N = \begin{pmatrix} 0 \\ 1 \end{pmatrix}$
		$J_{\frac{1}{4}} = \begin{pmatrix} \exp(i\pi/4) & 0 \\ 0 & \exp(i3\pi/4) \end{pmatrix}$
	Unknown:	e_I^N
Relevant equation		$e_T^N = J_{\frac{1}{4}}\, e_I^N$

$$
\begin{aligned}
e_T^N &= \begin{pmatrix} \exp(i\pi/4) & 0 \\ 0 & \exp(i3\pi/4) \end{pmatrix} \begin{pmatrix} 0 \\ 1 \end{pmatrix} \\
&= \begin{pmatrix} 0 \\ \exp(i3\pi/4) \end{pmatrix} \\
&= \exp(i3\pi/4) \begin{pmatrix} 0 \\ 1 \end{pmatrix}
\end{aligned}
$$

The transmitted wave is still plane-polarised in the vertical plane. It has the same amplitude as the incident wave but has suffered a phase delay of $3\pi/4$.

Example 5.16

Two quarter-wave plates, with fast axes vertical, are placed in series. The incident beam is plane-polarised at 45° to the axes. Use the method of matrices to determine the state of polarisation of the transmitted beam.

The normalised Jones vector for the transmitted beam is found by pre-multiplying the Jones vector for the incident beam by the Jones matrices for the two quarter-wave plates.

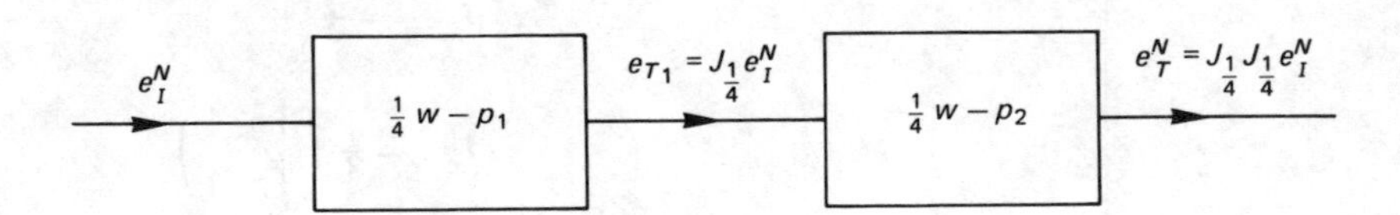

Data	*Given:*	$e_I^N = \frac{1}{\sqrt{2}} \begin{pmatrix} 1 \\ 1 \end{pmatrix}$
		$J_{\frac{1}{4}} = \begin{pmatrix} \exp(i\pi/4) & 0 \\ 0 & \exp(-i\pi/4) \end{pmatrix}$
	Unknown:	e_T^N
Relevant equation		$e_T^N = J_{\frac{1}{4}}\, J_{\frac{1}{4}}\, e_I^N$

Substitution gives

$$
e_T^N = \frac{1}{\sqrt{2}} \begin{pmatrix} \exp(i\pi/4) & 0 \\ 0 & \exp(-i\pi/4) \end{pmatrix} \begin{pmatrix} \exp(i\pi/4) & 0 \\ 0 & \exp(-i\pi/4) \end{pmatrix} \begin{pmatrix} 1 \\ 1 \end{pmatrix} \qquad (1)
$$

$$
= \frac{1}{\sqrt{2}} \begin{pmatrix} \exp(i\pi/4) & 0 \\ 0 & \exp(-i\pi/4) \end{pmatrix} \begin{pmatrix} \exp(i\pi/4) \\ \exp(-i\pi/4) \end{pmatrix}
$$

$$= \frac{1}{\sqrt{2}} \begin{pmatrix} \exp(i\pi/2) \\ \exp(-i\pi/2) \end{pmatrix}$$

$$= \frac{1}{\sqrt{2}} \begin{pmatrix} i \\ -i \end{pmatrix}$$

$$= \frac{i}{\sqrt{2}} \begin{pmatrix} 1 \\ -1 \end{pmatrix} \qquad (2)$$

This represents a plane-polarised wave inclined at 135° to the slow axes of the wave plates.

Each quarter-wave plate retards the Y-component wave by $\pi/2$, thus producing an overall phase delay of π. This means that the plane of polarisation of the incident wave has been 'flipped-over' to the other diagonal of the amplitude rectangle, as shown in Fig. 5.4(c).

Example 5.17

Vertically polarised light is incident on a polaroid with the transmission axis set at 45° to the positive Y-axis of a right-handed Cartesian coordinate system. The transmitted light then passes through a right-handed circular polariser. Using Jones matrices, determine the state of polarisation of the transmitted light.

As in Example 5.16, the Jones vector for the incident light is pre-multiplied by the Jones matrices for the linear polariser and the circular polariser.

Data	*Given:*	$e_I^N = \begin{pmatrix} 0 \\ 1 \end{pmatrix}$
		$J_{\text{cir R}} = \frac{1}{2} \begin{pmatrix} 1 & i \\ -i & 1 \end{pmatrix}$
		$J_{\text{pol}} = \frac{1}{2} \begin{pmatrix} 1 & 1 \\ 1 & 1 \end{pmatrix}$
	Unknown:	e_T^N
Relevant equation		$e_T^N = J_{\text{cir R}} \times J_{\text{pol}} \times e_I^N$

$$e_T^N = \tfrac{1}{4} \begin{pmatrix} 1 & i \\ -i & 1 \end{pmatrix} \begin{pmatrix} 1 & 1 \\ 1 & 1 \end{pmatrix} \begin{pmatrix} 0 \\ 1 \end{pmatrix}$$

$$= \tfrac{1}{4} \begin{pmatrix} 1 & i \\ -i & 1 \end{pmatrix} \begin{pmatrix} 1 \\ 1 \end{pmatrix}$$

$$= \tfrac{1}{4} \begin{pmatrix} 1 + i \\ 1 - i \end{pmatrix} \qquad (1)$$

As

$$\left. \begin{aligned} 1 + i &= \sqrt{2} \exp(i\pi/4) \\ &\text{and} \\ 1 - i &= \sqrt{2} \exp(-i\pi/4) \end{aligned} \right\} \qquad (2)$$

we can re-write (1) as

$$e_T^N = \frac{\sqrt{2}}{4}\begin{pmatrix}\exp(i\pi/4)\\ \exp(-i\pi/4)\end{pmatrix}$$

$$= \frac{\sqrt{2}}{4}\exp(i\pi/4)\begin{pmatrix}1\\ \exp(-i\pi/2)\end{pmatrix}$$

$$= \frac{\sqrt{2}}{4}\exp(i\pi/4)\begin{pmatrix}1\\ -i\end{pmatrix} \tag{3}$$

Thus the transmitted light is right-handed circularly polarised. The normalised irradiance is $(\sqrt{2}/4)^2$, i.e. $\frac{1}{8}$.

5.10 Questions

5.1

Two sheets of polaroid have their transmission axes parallel and oriented vertically. A third sheet is inserted between them with its transmission axis at 30° to the vertical. If the combination is illuminated at normal incidence with unpolarised light of irradiance I_o determine the irradiance of the emerging light.

5.2

A half-wave plate of quartz is 30 μm thick. Quasi-monochromatic light of wavelength 589.0 nm is incident normally on the plate. Calculate the transmission time for the O-wave if the refractive index of quartz for the E-wave is 1.553 and the speed of light is taken as 2.98×10^8 m s^{-1}.

5.3

A beam of plane-polarised He–Ne laser light (633 nm) is incident normally on a calcite crystal. Determine the thickness of the crystal which produces a phase difference of $\pi/6$ between the O- and E-waves. State precisely the orientations of the optic axis relative to the plane of incidence and the crystal faces. Take $n_E = 1.486$ and $n_O = 1.658$.

5.4

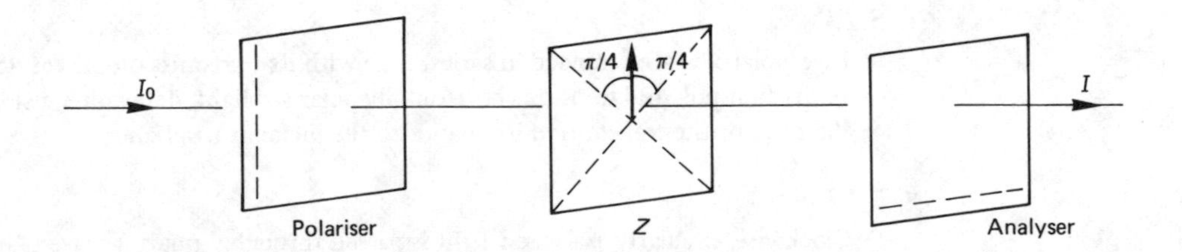

The fast and slow axes of the birefringent plate Z are held at an angle of $\pi/4$ to the vertical. Find the irradiance ratio I/I_o if Z is (a) a quarter-wave plate; (b) a half-wave plate.

5.5

Two polaroid sheets with transmission axes parallel are placed on each side of a piece of quartz cut normal to its optic axis. Given that the specific rotatory power of quartz is 21.7° mm^{-1} for a wavelength of 589.3 nm, determine the thickness of quartz required to produce the maximum transmitted irradiance.

Calculate the thickness required for a wavelength of 400 nm.

5.6

Write equations to describe the following polarised waves:

(i) A plane-polarised wave with optical vector lying at an angle of $2\pi/3$ with the positive Y-axis.

(ii) A left-handed circularly polarised wave.

(iii) A right-handed elliptically polarised wave with major axis lying along the Y-axis and the ratio of the major to minor axis equal to 3.

(iv) A left-handed elliptically polarised wave with major axis lying along the Z-axis and the ratio of major to minor axes equal to 2.

5.7

Determine the Brewster angle for flint glass ($n = 1.650$) immersed in oil ($n = 1.800$).

5.8

An electric field is increased continuously across a Kerr cell containing nitrobenzene until the transmitted irradiance has gone through two maxima. Determine the electrode separation of the cell used if the incident light passes through 3.0 cm of nitrobenzene and the potential difference between the electrodes is 3.0 kV. Take the electro-optic constant to be 2.4×10^{-12} V^{-2} m.

5.9

Use the experimental arrangement of Example 5.12 but replace Q_1 and Q_2 with half-wave plates. Are fringes observed?

5.10

Q_1 is a half-wave plate and Q_2 is a quarter-wave plate. Are fringes still observed as the polaroid sheet P is rotated?

5.11

Describe the state of polarisation of the wave:

$$\mathbf{E}_Y = -\hat{\mathbf{j}} A \sin (kx - \omega t)$$

$$\mathbf{E}_Z = \hat{\mathbf{k}}\, 2A \sin \left(kx - \omega t + \frac{\pi}{2} \right)$$

5.12

Use Jones matrices to show that a combination of a right-handed and a left-handed circular polariser is a complete absorber.

5.13

Five polaroids are arranged in series, each with its transmission axis at 45° to its predecessor. If the first polaroid transmits horizontally polarised light, determine, using the Jones calculus, the ratio of the transmitted irradiance to the incident irradiance.

5.14

Clockwise circularly polarised light is passed through a quarter-wave plate having its fast-axis vertical. Determine the state of polarisation of the transmitted light.

5.15

Brewster's law indicates that the polarising angle depends on:

(a) the frequency of the incident light;

(b) the ratio of the speed of light in air and the dielectric;

(c) the reflection coefficient of the dielectric surface;

(d) the angle between the incident and the refracted light;

(e) none of these but . . .

5.16

The transmission axis of polaroid lies at right angles to the chains of polyvinyl alcohol molecules because:
(a) electrons cannot move far along the molecular chains;
(b) electrons in a chain cannot move far in a direction at right angles to the chain;
(c) positive ions move very little from their equilibrium positions;
(d) the presence of iodine absorbs the incident radiation;
(e) none of these but . . .

5.17

If a partially polarised beam is composed of 3 W m^{-2} of plane-polarised light and 7 W m^{-2} of natural light, the degree of polarisation is:
(a) 3;
(b) 0.3 W m^{-2};
(c) 300;
(d) 0.43;
(e) none of these but . . .

5.18

Huyghens's principle says that:
(a) wavefronts are spherical in optically homogeneous materials;
(b) wavefronts are spherical in uniaxial crystals;
(c) the shape of a wavefront depends on the refractive index;
(d) every point on a wavefront acts as a source of secondary wavelets and the envelope of all the wavelets corresponds to the primary wave at that time;
(e) none of these but . . .

5.19

If a substance has a large Kerr constant then it means that:
(a) the molecules are light;
(b) the molecules are heavy;
(c) the molecules do not polarise very easily;
(d) the molecules polarise very easily;
(e) none of these but . . .

5.20

In photo-elastic stress analysis the isochromatic fringes are those for which the difference between the principal stresses gives an optical path difference of:
(a) zero;
(b) $\lambda/2$;
(c) $m\lambda$; (m = integer)
(d) $\lambda/4$;
(e) none of these but . . .

5.21

In a Young's slits experiment, for interference to occur with incident plane-polarised light the latter must have:
(a) its plane of polarisation at right angles to both slits;
(b) its plane of polarisation parallel to one slit and at right angles to the other;
(c) its plane of polarisation parallel to both slits;
(d) a high degree of monochromaticity;
(e) none of these but . . .

5.22

The optic axis may be defined as:
(a) a direction of rotational symmetry;
(b) a direction along which the O- and E-rays always travel with the same speed;
(c) the direction along which the O- and E-rays have their greatest speeds;
(d) the direction along which the O- and E-rays have their least speeds;
(e) none of these but . . .

5.23

Two sheets of polaroid are inadequate to use in an optical polarimeter because the optical rotation θ:
(a) is always too small;
(b) sometimes too large;
(c) may be very small;
(d) may be greater than 2π;
(e) none of these but . . .

5.24

Crystalline quartz is optically active because:
(a) it is trigonal;
(b) its atoms are arranged in a clockwise helix;
(c) its atoms are arranged in an anti-clockwise helix;
(d) thin mica plates cut parallel to the optic axis and twisted clockwise rotate the plane of polarisation;
(e) none of these but . . .

5.11 Answers to Questions

5.1 $9I_o/32$
Use Malus's law, remembering that the transmitted irradiance by polaroid 1 (polariser) is $I_o/2$.

5.2 1.55×10^{-13} s
Use: $d\,(n_E - n_O) = \lambda/2$

5.3 308 nm
$2\pi d\,(n_O - n_E)/\lambda = \pi/6$. Hence find d.

5.4 (a) 1/8
(b) 1/2
The amplitude of the light incident on the crystal is $(I_o/2k)^{\frac{1}{2}}$, where k is a constant of proportionality. Component amplitudes along the fast and slow axes are $A/\sqrt{2}$ and components parallel to the transmission axis of the analyser are $A/2$.

5.5 8.3 mm
The maximum transmitted irradiance occurs when plane-polarised light produced by the polariser is rotated through 180°. The length of quartz required is 180/21.7 mm.
3.8 mm
The specific rotatory power is defined for yellow light. It varies as λ^{-2}.

5.6 (i) $E_Y = A \cos(kx - \omega t)$

$E_Z = A\sqrt{3} \cos(kx - \omega t + \pi) = -A\sqrt{3} \sin(kx - \omega t)$

(ii) $E_Y = A \cos(kx - \omega t)$

$$E_Z = A \cos\left(kx - \omega t + \frac{\pi}{2}\right) = -A \sin(kx - \omega t)$$

(iii) $E_Y = 3A \cos(kx - \omega t)$

$$E_Z = A \cos\left(kx - \omega t - \frac{\pi}{2}\right) = A \sin(kx - \omega t)$$

(iv) $E_Y = A \cos(kx - \omega t)$

$$E_Z = 2A \cos\left(kx - \omega t + \frac{\pi}{2}\right) = -2A \sin(kx - \omega t)$$

5.7 42°30′

The refractive indices given are with respect to air. You require the refractive index of the glass with respect to oil, i.e. 1.650/1.800.

5.8 6.6×10^{-1} mm

The effect of the electric field is to introduce a phase difference of 3π, as the transmitted irradiance has undergone two maxima. Hence

$$2\pi K V^2 L/d^2 = 3\pi$$

from which the electrode separation d can be calculated.

5.9 Yes. The light emerging from each half-wave plate will be plane-polarised, and oriented along the same direction, whatever the direction of the transmission axis of the polaroid.

5.10 Only when the transmission axis of the polaroid is parallel to, or almost parallel to, the crystal axes.

5.11 Convert the wave equations to:

$$\mathbf{E}_Y = \hat{\mathbf{j}} A \cos\left(kx - \omega t + \frac{\pi}{2}\right)$$

$$\mathbf{E}_Z = \hat{\mathbf{k}}\, 2A \cos(kx - \omega t)$$

Now it can be seen that the Z-wave leads the Y-wave by $\pi/2$. As the amplitudes are unequal, the resultant wave is clockwise-elliptically polarised light having its major axis lying along the Z-axis.

5.12 $$J_{\text{comb.}} = \tfrac{1}{4}\begin{pmatrix} 1 & \mathrm{i} \\ -\mathrm{i} & 1 \end{pmatrix}\begin{pmatrix} 1 & -\mathrm{i} \\ \mathrm{i} & 1 \end{pmatrix}$$

$$= \tfrac{1}{4}\begin{pmatrix} 0 & 0 \\ 0 & 0 \end{pmatrix}$$

5.13 $$e_T^N = \tfrac{1}{4}\begin{pmatrix} 1 & 0 \\ 0 & 0 \end{pmatrix}\begin{pmatrix} 1 & -1 \\ -1 & 1 \end{pmatrix}\begin{pmatrix} 0 & 0 \\ 0 & 1 \end{pmatrix}\begin{pmatrix} 1 & 1 \\ 1 & 1 \end{pmatrix}\begin{pmatrix} 1 \\ 0 \end{pmatrix}$$

$$= \tfrac{1}{4}\begin{pmatrix} -1 \\ 0 \end{pmatrix}$$

This is horizontally polarised light. The irradiance ratio is $(\frac{1}{4})^2$, i.e. $\frac{1}{16}$. You can check this result using Malus's law. The state of polarisation of the incident light is not required because the first polariser produces a horizontally polarised beam, which acts as the incident beam for the second polariser.

5.14 $$e_T^N = \frac{1}{\sqrt{2}}\begin{pmatrix}\exp(i\pi/4) & 0\\ 0 & \exp(-i\pi/4)\end{pmatrix}\begin{pmatrix}1\\ -i\end{pmatrix}$$

$$= \frac{1}{\sqrt{2}}\begin{pmatrix}\exp(i\pi/4)\\ \exp(-i2\pi/4)\end{pmatrix}$$

$$= \frac{1}{\sqrt{2}}\exp(i\pi/4)\begin{pmatrix}1\\ \exp(-i\pi)\end{pmatrix}$$

$$= \frac{1}{\sqrt{2}}\exp(i\pi/4)\begin{pmatrix}1\\ -1\end{pmatrix}$$

This is plane-polarised light vibrating at 135° to the positive Y-axis.

5.15 (a)
(b) is a special case only.

5.16 (b)

5.17 (e)
The degree of polarisation is defined by $I_{pp}/(I_{pp} + I_{nat})$ and equals 0.3.

5.18 (e)
See Question 4.12. Although (a) is correct it is a consequence of Huyghens' principle.

5.19 (d)

5.20 (c)

5.21 (c)

5.22 (a)

5.23 (c)

5.24 (b); (c)
The O- and E-wavefronts do not touch along the optic axis direction. Quartz can exist in a right-handed and a left-handed form. The silicon and oxygen atoms are arranged in a helix with the axis along the optic axis direction.

Appendix

Consider the following simultaneous equations:

$$m_1 = ax_1 + by_1 \qquad \text{(A1)}$$
$$m_2 = cx_2 + dy_2$$

These can be written in a shorthand way as

$$\begin{pmatrix} m_1 \\ m_2 \end{pmatrix} = \begin{pmatrix} a & b \\ c & d \end{pmatrix} \begin{pmatrix} x_1 \\ y_1 \end{pmatrix} \qquad \text{(A2)}$$

There are two column matrices

$$\begin{pmatrix} m_1 \\ m_2 \end{pmatrix} \text{ and } \begin{pmatrix} x_1 \\ y_1 \end{pmatrix}$$

which are usually referred to as column vectors. The central matrix is a two-row × two-column matrix (or 2 × 2 matrix for short). Call it M_1.

Consider two further simultaneous equations:

$$n_1 = fx_1 + gy_1 \qquad \text{(A3)}$$
$$n_2 = hx_1 + jy_1$$

which we can write in matrix form as

$$\begin{pmatrix} n_1 \\ n_2 \end{pmatrix} = \begin{pmatrix} f & g \\ h & j \end{pmatrix} \begin{pmatrix} x_1 \\ y_1 \end{pmatrix} \qquad \text{(A4)}$$

in which

$$N_1 = \begin{pmatrix} f & g \\ h & j \end{pmatrix}$$

Let us put:

$$p_1 = m_1 + n_1$$

and

$$p_2 = m_2 + n_2$$

Then

$$p_1 = (a + f)x_1 + (b + g)y_1$$

and $\qquad$ (A5)

$$p_2 = (c + h)x_1 + (d + j)y_1$$

or

$$\begin{pmatrix} p_1 \\ p_2 \end{pmatrix} = \begin{pmatrix} a + f & b + g \\ c + h & d + j \end{pmatrix} \begin{pmatrix} x_1 \\ y_1 \end{pmatrix} \qquad \text{(A6)}$$

This new 2×2 matrix is formed by adding the individual elements of M_1 and N_1 so that

$$\begin{pmatrix} a+f & b+g \\ c+h & d+j \end{pmatrix} = \begin{pmatrix} a & b \\ c & d \end{pmatrix} + \begin{pmatrix} f & g \\ h & j \end{pmatrix}$$

or

$$M_1 + N_1 = P_1 \tag{A7}$$

Subtraction of matrices can be carried out in a similar way by subtracting individual elements.

Now suppose that we have the following sets of equations:

$$\begin{aligned} y_1 &= ax_1 + bx_2 \\ y_2 &= cx_1 + dx_2 \end{aligned} \quad \text{or} \quad \begin{pmatrix} y_1 \\ y_2 \end{pmatrix} = \begin{pmatrix} a & b \\ c & d \end{pmatrix} \begin{pmatrix} x_1 \\ x_2 \end{pmatrix} \tag{A8}$$

and

$$\begin{aligned} z_1 &= fy_1 + gy_2 \\ z_2 &= hy_1 + jy_2 \end{aligned} \quad \text{or} \quad \begin{pmatrix} z_1 \\ z_2 \end{pmatrix} = \begin{pmatrix} f & g \\ h & j \end{pmatrix} \begin{pmatrix} y_1 \\ y_2 \end{pmatrix} \tag{A9}$$

Then

$$\begin{aligned} z_1 &= f(ax_1 + bx_2) + g(cx_1 + dx_2) = (af + cg)x_1 + (bf + dg)x_2 \\ z_2 &= h(ax_1 + bx_2) + j(cx_1 + dx_2) = (ah + cj)x_1 + (bh + dj)x_2 \end{aligned} \tag{A10}$$

In matrix form (A10) is

$$\begin{pmatrix} z_1 \\ z_2 \end{pmatrix} = \begin{pmatrix} af + cg & bf + dg \\ ah + jc & bh + dj \end{pmatrix} \begin{pmatrix} x_1 \\ x_2 \end{pmatrix} \tag{A11}$$

So we see that

$$\begin{pmatrix} f & g \\ h & j \end{pmatrix} \begin{pmatrix} a & b \\ c & d \end{pmatrix} = \begin{pmatrix} fa + gc & fb + gd \\ ha + jc & hb + jd \end{pmatrix} \tag{A12}$$

Each element in the matrix is the product of a row vector and a column vector. For example the first element is obtained from

$$(f \quad g) \begin{pmatrix} a \\ c \end{pmatrix}$$

(A12) summarises the method for multiplying matrices. It can be extended to any number of matrices. The one point to bear in mind is the ordering of the matrices, e.g.

$$\begin{pmatrix} f & g \\ h & j \end{pmatrix} \begin{pmatrix} a & b \\ c & d \end{pmatrix} \neq \begin{pmatrix} a & b \\ c & d \end{pmatrix} \begin{pmatrix} f & g \\ h & j \end{pmatrix}$$

The multiplication is said to be *non-commutative*.

Bibliography

Born, M. and Wolf, E. (1965). *Principles of Optics*, Pergamon Press, Oxford

Ditchburn, R. W. (1963). *Light*, Interscience, London

Gerrard, A. and Burch, J. M. (1975). *Introduction to Matrix Methods in Optics*, John Wiley, London

Hecht, E. and Zajac, A. (1974). *Optics*, Addison-Wesley, London

Jenkins, F. A. and White, H. E. (1976). *Fundamentals of Optics*, McGraw-Hill Kogakusha, London

Klein, M. V. (1970). *Optics*, John Wiley, London

Longhurst, R. S. (1957). *Geometrical and Physical Optics*, Longman Green, London

Meyer-Arendt, J. R. (1972). *Introduction to Classical and Modern Optics*, Prentice-Hall, New Jersey

Nussbaum, A. (1976). *Contemporary Optics for Scientists and Engineers*, Prentice-Hall, New Jersey

Nussbaum, A. (1968). *Geometric Optics: An Introduction*, Addison-Wesley, New York

Tenquist, D. W., Whittle, R. M. and Yarwood, J. (1969). *University Optics*, Iliffe, London

Young, A. D. (1976). *Fundamentals of Waves, Optics and Modern Physics*, McGraw-Hill, London

Young, M. (1986). *Optics and Lasers*, Springer-Verlag, Berlin

Index